SOLUTIONS MANUAL

TO ACCOMPANY

THE WORLD
OF THE CELL

FIFTH EDITION

Wayne M. Becker
University of Wisconsin, Madison

Benjamin
Cummings

San Francisco Boston New York
Capetown Hong Kong London Madrid Mexico City
Montreal Munich Paris Singapore Sydney Tokyo Toronto

Sponsoring Editor: Michele Sordi
Publishing Assistant: Michael J. McArdle
Composition: The Left Coast Group
Cover Design: Tony Asaro

ISBN 0-8053-4856-5

Benjamin
Cummings

1 2 3 4 5 6 7 8 9 10—VHG—06 05 04 03 02

Contents

Preface

This *Solutions Manual* provides solutions for all of the problems that appear at the end of each chapter in the Fifth Edition of *The World of the Cell.* The inclusion of problem sets in the text reflects our conviction that one learns science not just by reading or hearing about it but by working with it. These solutions are meant to confirm your understanding when you arrive at the correct answer and to provide guidance in the appropriate problem-solving skills when you do not.

I am deeply indebted to Lewis Kleinsmith and Jeffrey Harden, co-authors of *The World of the Cell,* and to John Raasch, contributor of Chapters 12 and 15, for providing solutions to the problems that they wrote or revised for the Fifth Edition. I retain responsibility for any errors of omission or commission, and I welcome feedback from users of this manual, students and instructors alike. Please address comments and suggestions directly to me.

Wayne M. Becker
Department of Botany B115 Birge Hall
University of Wisconsin–Madison
Madison, Wisconsin 53706
wbecker@facstaff.wisc.edu

1

The World of the Cell: A Preview

1-1. (a) C   (d) B   (g) G   (j) G
     (b) B   (e) G   (h) B, C (k) C
     (c) G   (f) C   (i) B   (l) B

1-2. (a) Bacterial cell: $V = \pi r^2 h = (3.14)(0.5)^2(2.0) = $ **1.57 μm^3**.

     Liver cell: $V = 4\pi r^3/3 = 4(3.14)(10)^3/3 = $ **4200 μm^3**.

     Palisade cell: $V = \pi r^2 h = (3.14)(10)^2(35) = $ **11,000 μm^3**.

     (b) Bacterial cells in a liver cell: $4200/1.57 = $ **2700**.

     (c) Liver cells in a palisade cell: $11,000/4200 = $ **2.62**.

1-3. (a) Light microscope: limit of resolution = 200 nm. Because 1 membrane has a thickness of about 8 nm, the number of membranes that must be aligned laterally is $200/8 = $ **25 membranes.**

     Electron microscope: limit of resolution = 0.1-0.2 nm. With a thickness of about 8 nm, a single membrane can be seen using an electron microscope. (To contrast light and electron microscopy in terms of resolving power, compare parts (a) and (b) of Figure 1-3 in the textbook, looking specifically at the nuclear membranes.)

     b) Liver cell: $V = 4200\ \mu m^3$ (from Problem 1-2a).

     Ribosome: $V = 4\pi r^3/3 = 4(3.14)(0.0125)^3/3 = 8.2 \times 10^{-6}\ \mu m^3$.

     Ribosomes in a liver cell: $4200/(8.2 \times 10^{-6}) = $ **5×10^8**.

     (c) Bacterial cell: $V = 1.57\ \mu m^3$ (from Problem 1-2a).

     DNA molecule: $V = \pi r^2 h = (3.14)(1\ nm)^2(1.36\ mm)$

     $= (3.14)(0.001\ \mu m)^2(1.36 \times 10^3\ \mu m)$

     $= (3.14)(1 \times 10^{-6}\ \mu m^2)(1.36 \times 10^3\ \mu m)$

     $= (3.14)(1.36)(10^{-3})\ \mu m^3 = 4.3 \times 10^{-3}\ \mu m^3$.

     DNA $= (4.3 \times 10^{-3})\ /1.57 = 0.0027 = $ **0.27% of cell volume.**

1-4.   (a)   The *limit of resolution* of a microscope is a measure of how close together two points can be and still be distinguished from one another when viewed through the microscope. (Note that the limit of resolution is related inversely to magnification; the greater the magnification of a microscope, the smaller its limit of resolution will be.) Hooke's microscope could only magnify objects 30-fold, so its limit of resolution was one-thirtieth that of the human eye: 0.25 mm/30 = 0.0083 mm, or 8.3 μm. Van Leeuwenhoek's microscope could magnify objects 300-fold, which was ten times greater magnification than that of Hooke's microscope, so the limit of resolution of van Leeuwenhoek's microscope was one-tenth that of Hooke's microscope: 8.3 μm/10 = 0.83 μm.

(b)   The smallest structures that Hooke was able to see were about 8.3 mm in one dimension, which would have allowed him to see the plant and animal cells shown in Figure 1A-1 of the textbook but not a typical bacterial cell.

(c)   The smallest structures that van Leeuwenhoek was able to see were about 0.83 μm in one dimension, which would have allowed him to see all three of the structures shown in Figure 1A-1.

(d)   The limit of resolution of a modern light microscope is about 200-350 nm (0.2-0.35 μm) in one dimension, so structures must be at least this large in two dimensions to be visualized.

(e)   The seven structures could be visualized by the indicated microscopes (H = Hooke's microscope; V = van Leeuwenhoek's microscope; C = contemporary light microscope):

Mycoplasma (d = 0.3 μm): C

Liver cell (d = 20 μm): H, V, C

Nucleus (d = 6 μm): V, C

Chloroplast (2 μm $\times$ 8 μm): V, C

Mitochondrion (1 μm $\times$ 2 μm): V, C

Peroxisome (d = 0.5 μm): C

Microtubule (25 nm $\times$ 1 μm): possibly C; most likely none

1-5.   (a)   Resolution is limited primarily by the wavelength of the photons or electrons that are used to visualize the structures of interest.

(b)   The limit of resolution, r, is the minimum distance between two objects at which they can still be resolved from each other. The value of r is directly proportional to wavelength, so ultraviolet light with wavelengths in the range 200-300 nm has a limit of resolution about half that of visible light, for which the wavelength range is 400-700 nm. This means that two objects that need to be separated by a distance d in order to be resolved from each other using visible light can be identified as two separate objects at a distance $d/2$ when ultraviolet light is used. (For details, see equations A-1 through A-4 on pages 818–819 of the textbook Appendix.)

(c)   From Figure 1-2 in the textbook, the limits of resolution (*i.e.*, the bottoms of the arrows) for the unaided eye and the light microscope are about 0.2-0.3 μm (200-300 nm) and 0.2-0.3 nm, respectively. The upper limit of magnification using a light microscope is therefore about 1000-fold. This is called "useful magnification" because up to this limit, increases in magnification make it possible to resolve points that are closer and closer together.

(d)   Beyond the limit of useful magnification, further magnification will not enable the viewer to see points that are still closer together because the limit of resolution is set by the wavelength of the light. Any enlargement of the image beyond the useful magnification of the optical system is therefore called "empty magnification" because it provides no additional resolution and hence, no additional information about the object(s) being studied.

1-6.   (a)   Initially thought to be true because animal cells do not have cell walls, which made it hard to distinguish individual cells using the crude microscopes available to early investigators; shown by Schwann (1839) to be incorrect for cartilage cells, which have well-defined boundaries of collagen fibers, and later extended to all animal cells.

(b)   Initially thought to be true because living organisms seem to increase in complexity spontaneously, unlike other systems known to early chemists or physicists; misconception laid to rest by Wöhler's demonstration (1828) that urea, a compound made by living organisms, could be synthesized in the laboratory from an inorganic starting compound.

(c)   Originally thought to be true because the order of nucleotide monomers in DNA was erroneously considered to be an invariant tetranucleotide repeating sequence; disproved by Avery et al. for bacteria (1944) and by Hershey and Chase for bacterial viruses (1952).

(d)   Initially thought to be true because of the demonstration by Pasteur that yeast cells were needed for alcoholic fermentation; Buchner and Buchner showed later (1897) that extracts from yeast cells could substitute for intact cells, an effect we now know to be due to the presence in the extracts of the enzymes that catalyze the various reactions in the fermentation process.

1-7.   (a)   Consistent with early electron microscopy in which membranes appeared as two parallel electron-dense lines (thought to be the outer protein layer) separated by an unstained space (thought to be the lipid interior of the membrane); disproved by the demonstration that membrane proteins are globular structures located within the membrane, not just on its surface.

(b)   Postulated because of the early demonstration that high-energy intermediates are involved in ATP generation during the glycolytic pathway; now known not to be the case for ATP generation by mitochondria, where the driving force turns out to be the energy of an electrochemical proton gradient, with no high-energy phosphorylated intermediates involved.

(c)   Shown to be true for the photosynthetic algae used for the first studies of ^{14}C fixation; now understood to be true only for so-called "C_3 plants," because "C_4 plants" are known in which the initial product of carbon fixation is a four-carbon compound instead.

(d)   Believed to be universally true until the discovery of parasites called trypanosomes in which the enzymes of the glycolytic pathway responsible for the conversion of sugar into pyruvate are compartmentalized into a membrane-bounded organelle called a *glycosome*.

(e)   Thought to be universally true until the discovery of Z-DNA, in which the two strands form a left-handed helix.

(f)   Thought to be universally true until the discovery that the genetic code used in mitochondria is different in several aspects from the code that appears to operate everywhere else.

1-8.   (a)   The four scientists might have found that they had common interests in the origin of cells (from preexisting cells, not by "spontaneous generation") and the nature of the genetic information that organisms pass on to their progeny. (Bear in mind, however, that we are answering the question from the perspective of almost 125 years of further discoveries and insights. In 1875, the four scientists might with equal likelihood have concluded that they had few, if any, interests in common.)

(b)   Given Pasteur's demonstration that living, preexisting yeast cells are needed to carry out the fermentation of sugar into alcohol, he would almost certainly have been interested in Virchow's conclusion that all cells come from preexisting cells. (In fact, he would most likely have found Virchow's addition to the cell theory of Schleiden and Schwann particularly encouraging in light of the earlier belief, debunked by Pasteur's work, that yeast cells arose by "spontaneous generation" in the presence of such substrates as fermentable grape juice.)

(c)   In light of Virchow's conclusion that cells (and therefore organisms as well, presumably) arise only from preexisting cells, he would likely have been very interested in Mendel's evidence that traits such as those he observed in pea plants were heritable and could be passed from one generation to the next by transmission of "hereditary factors," because the genetic continuity between generations implied by Mendel's findings would have fit well with the physical continuity of generations inherent in Virchow's contribution to cell theory.

(d)   If Miescher were sufficiently foresighted, he might have been most interested in the work of Mendel because Miescher's discovery of what we now call DNA was eventually to provide the chemical basis for the heritability of traits described by Mendel.

1-9.   White presumably means that the "truth" represented by a scientific "fact" is never a dogmatic, once-and-for-all statement that explains a phenomenon absolutely or definitively but is instead no more than a reasonably plausible interpretation of the current data, tenable only until a better, more plausible explanation is presented. This concept of "truth" applies aptly to scientific "facts" because a given scientific fact is accepted by the scientific community only until a better explanation is offered, based on more convincing observations and/or experimental data. White's statement relates to the scientific method shown in Figure 1B-1 of the textbook because the intention of a scientific experiment is to test predictions based on currently accepted explanations of a specific phenomenon and thereby to determine whether those predictions, and hence the current explanation, are substantiated by the new experimental evidence or are in need of reinterpretation or revision.

1-10.   (a)   The experimental process should begin by testing the hypothesis that the heartburn is due to the pizza. Rather than simply "keeping track of your eating habits for a few weeks," which is observational biology, you need to design an experimental protocol whereby you can test the hypothesis by deliberately consuming pizza on certain nights and refraining from pizza on others. (Ideally, such a protocol should ensure that the subject is not aware of which evenings pizza is consumed, to avoid the so-called "placebo effect" in which heartburn might be induced just by *knowing*—or *thinking*—that you've eaten pizza. But given the nature of pizza, it might be difficult to build this control into the experiment!) If you find a good correlation between pizza consumption and heartburn and are able to confirm this repeatedly—note the requirement for repeatable observations—it seems reasonable to conclude that your heartburn is, in fact, caused by eating pizza. Having tentatively confirmed that hypothesis, you are now ready to test the further hypothesis that the heartburn is caused specifically by one or more of the suspected ingredients. To do so, you need to design an experimental protocol whereby each ingredient is tested systemacally, both alone and in combination with one or both of the other ingredients. In addition, you will need a "control," consisting of plain pizza with none of the suspected ingredients. (A total of seven possible "treatments" will be involved in your experimental design; can you specify the seven?) Again, repeatability of the observations is an important criterion before you can conclude that one, two, or all three ingredients cause heartburn, either individually or in specific combinations (or, alternatively, that none of the ingredients appears to be causally involved).

(b)   If you compare the above process with the scientific method as illustrated in Figure 1B-1 of the textbook, you should be able to convince yourself that each of the steps outlined there is present in the scenario described in part (a), including the reference to prior knowledge represented by the initial correlation noted between heartburn and consumption of pizza with all three ingredients.

CHAPTER

2

The Chemistry of the Cell

2-1.   (a)   Carbon is a smaller atom than silicon (atomic weight of 12 instead of 28). It therefore forms especially stable covalent bonds, because the strength of a covalent bond is inversely proportional to the atomic weights of the elements involved.

      (b)   Carbon combines with oxygen to form simple gaseous molecules and is therefore not removed from circulation in an aerobic environment, as occurs with insoluble silicon oxides.

      (c)   Carbon is one of only three elements (the other two are oxygen and nitrogen) that readily form strong multiple bonds. Carbon forms both double and triple bonds; the former especially are important in biological compounds.

      (d)   Carbon also has four valence electrons, but it is much less abundant in the earth's crust than silicon; carbon is not obviously fitter in this regard.

      (e)   Polymers of carbon are stable in water because of the strength of the carbon-carbon bond (see part a). They are also readily soluble if they have hydrophilic groups. Stability and solubility are important biologically because most of life involves an aqueous milieu.

2-2.   (a)   T; living organisms are essentially aqueous solutions containing many kinds of molecules, most of which are polar and hence readily soluble in water.

      (b)   T; oxygen is the ultimate electron acceptor in cellular respiration, with water as the product.

      (c)   F; the density of ice is less than that of water, thereby ensuring that ice will float on the surface of a body of water, where it will melt readily if the temperature of the surrounding air rises above the freezing point.

      (d)   T; this property explains the high specific heat and high heat of vaporization and hence the capacity of water to "buffer" cells and organisms against temperature changes.

      (e)   T; this property of water allows light to penetrate readily, such that submerged photosynthetic organisms (or parts of organisms) can receive sunlight.

      (f)   X; the lack of odor or taste is probably not a strategic advantage to most organisms.

(g)   T; high specific heat means that much heat is required to increase the temperature, which effectively "buffers" cells and organisms against temperature changes in response to changes in temperature of the environment.

(h)   T; high heat of vaporization means that much heat is required to convert water from a liquid to a gas, which means that organisms can be effectively cooled by evaporation of perspiration or other forms of liquid water from the skin or other surface of the organism.

2-3.   (a)   T

(b)   F; water has a higher heat of vaporization than most other liquids because of the hydrogen bonds that must be disrupted.

(c)   F; hydrophobic oil droplets in water coalesce not because of any intrinsic attraction of oil molecules for each other but because they have no affinity for polar molecules and therefore are not soluble in water.

(d)   T

(e)   F; the main driving force for protein folding is the tendency of hydrophobic side groups of amino acids to seek out an internal location where they will be shielded from the polar water molecules that surround the protein.

2-4.   (a) $E = 28,600/\lambda$; $\lambda = 28,600/E = 28,600/83 = $ **344 nm.**

(b)   Less stable; its cutoff comes at 409 nm.

(c)   More stable; it can only be broken by ultraviolet light with wavelengths less than 196 nm.

2-5.   (a)   Four; all combinations of alternate configurations at carbon atoms 2 and 3 (numbered from the carboxyl end of the molecule).

(b)   No stereoisomerism.

(c)   Two; both alternative configurations at carbon atom 2.

(d)   Four; all combinations of alternative configurations at carbon atoms 3 and 4.

2-6.   (a)   Hexadecane lacks a hydrophilic group and is therefore insoluble in an aqueous milieu. Palmitate, although otherwise identical to hexadecane in structure and chain length, has a hydrophilic carboxyl group, and is therefore amphipathic.

(b)   Compound ii; compound i is an ether and is not sufficiently polar to be soluble in an aqueous milieu, whereas compound ii has a hydrophilic terminus and is therefore soluble in water. (Compound ii is the amino acid isoleucine.)

2-7.   It is this asymmetry that renders water a polar molecule; most of the desirable properties of water as a solvent depend on this polarity.

2-8.   (a)   A single enzyme or set of enzymes can be used to add each successive monomeric unit.

(b)   Water is a readily available reactant in an essentially aqueous world.

2-9.   (a)   TMV virions self-assemble spontaneously without the input of energy or information, which means that all of the information necessary to direct their assembly must be already present in the RNA and/or proteins.

(b)   The strain-specific assembly of TMV in vivo is determined by the RNA, not the coat protein.

(c)   The information necessary to direct self-assembly of TMV virions appears to reside in the coat protein monomers.

(d)   The self-assembly of TMV virions is specific for TMV RNA.

(e)   The most stable configuration for TMV virions is achieved by the 3:1 ratio of nucleotides and coat protein monomers, and is therefore the product formed upon self-assembly regardless of the starting ratio of nucleotides and monomers.

2-10.   (a)   What is ultraviolet radiation?

(b)   What is the temperature-stabilizing capacity of water? (Or, alternatively, what is the specific heat of water?)

(c)   What is the activated form of a monomer?

(d)   What is the principle of self-assembly?

(e)   Why is the amphipathic nature of membrane components so important?

3

The Macromolecules of the Cell

(3-1.   (a)   2, 3, 6   (d)   1, 3, 9

     (b)   4, 7, 9, 10   (e)   4, 7, 9, 10

     (c)   4, 5, 9, 10   (f)   4, 5, 9, 10

3-2.

Bond	Amino Acids	Levels of Structure
Peptide (covalent)	All	Primary
Hydrogen	All	Secondary
Hydrophobic	Leucine	Tertiary, Quaternary
Ionic	Glutamate	Tertiary, Quaternary
Disulfide (covalent)	Cysteine	Tertiary, Quaternary

3-3.   (a)   Interior: valine, phenylalanine (hydrophobic).

        Exterior: aspartate, lysine (charged; hydrophilic).

        Either: glycine, alanine (small R groups; no strong affinity for or aversion to water).

     (b)   Alanine; phenylalanine; glutamate; methionine (the less polar member of each pair).

     (c)   The free sulfhydryl group is polar and ionizable; the disulfide bond is much less polar.

3-4.   The hydrophobic "patches" on the surfaces of the α and β subunits of hemoglobin represent sites of interaction between the subunits and as such are not on the surface of the intact tetrameric hemoglobin molecule, even though they are on the surface of the unassembled subunits.

3-5.   (a)   The amino acid glutamate is hydrophilic and ionizes at cellular pH, whereas valine is hydrophobic and nonionic. Substitution of the latter for the former is likely to change the chemical nature of that part of the molecule significantly.

     (b)   Aspartate is another acidic amino acid and is therefore a conservative change. Others that are unlikely to have major effects are the polar but uncharged amino acids serine, threonine, tyrosine and cysteine.

(c)   Yes, if the substitutions are always of like-for-like amino acids in terms of chemical properties. These are chemically conservative changes.

3-6.   (a)   To pull on both ends of an α-keratin polypeptide is to pull against the hydrogen bonds that account for its helical structure; these "give" readily, allowing the polypeptide to be stretched to its full, uncoiled length, at which point you would begin to pull against the covalent peptide bonds. For fibroin, you are pulling against the covalent peptide bonds immediately. (As an analogy, you might compare pulling on opposite ends of a coiled spring versus a straight length of uncoiled wire.)

(b)   Fibroin consists mainly of the two smallest amino acids, so it has few bulky R groups and can accommodate the constraints of a pleated sheet. Keratin, on the other hand, has most of the amino acids present, and the distance between bulky R groups is maximized when these protrude from a twisted helical shape.

3-7.   (a)   Hair proteins are first treated with a sulfhydryl reducing agent to break disulfide bonds and thereby destroy much of the natural tertiary structure and shape of the hair. After being "set" in the desired shape, the hair is treated with an oxidizing agent to allow disulfide bonds to re-form, but now between different cysteine groups, as determined by the positioning imposed by the curlers. These unnatural disulfide bonds then stabilize the desired configuration.

(b)   There are two reasons for the lack of permanence: (1) Disulfide bonds occasionally break and re-form spontaneously, allowing the hair proteins to return gradually to their original, thermodynamically more favorable shape. (2) Hair continues to grow, and the new α-keratin molecules will have the natural (correct) disulfide bonds.

(c)   There is probably a genetic difference in the positioning of cysteine groups and hence in the formation of disulfide bonds.

3-8.   (a)   k, f, d   (d)   none   (g)   k, d

(b)   k   (e)   k, f   (h)   k, f, d

(c)   f, d   (f)   f, d   (i)   f, d

3-9.   (a)   Circumference $= \dfrac{4 \times 10^6 \text{ pairs}}{\text{molecule}} \times \dfrac{1 \text{ turn}}{10 \text{ pairs}} \times \dfrac{3.4 \text{ nm}}{\text{turn}}$

$= 1.36 \times 10^6 \text{ nm} = \textbf{1.36 mm.}$

This poses a packaging problem, because the circumference of the DNA is about 680 times the length of the whole cell!

(b)   Circumference $= \pi d = (3.14)\,(1.8\ \mu m) = 5.65\ \mu m = \textbf{5650 nm.}$

$\dfrac{5650 \text{ nm}}{\text{molecule}} \times \dfrac{1 \text{ turn}}{3.4 \text{ nm}} \times \dfrac{10 \text{ pairs}}{\text{turn}} = \textbf{16,470 nucleotide pairs/molecule.}$

$16{,}470 \text{ pairs} \times \dfrac{1 \text{ polypeptide}}{1000 \text{ pairs}} = \textbf{16 polypeptides.}$

(c) $$\frac{2.8 \times 10^9 \text{ g}}{\text{mole}} \times \frac{1 \text{ mole}}{6.023 \times 10^{23} \text{ molecules}} \times \frac{1 \text{ molecule}}{4 \times 10^6 \text{ pair}} = \mathbf{1.16 \times 10^{-21}} \text{ **g/pair.**}$$

$$\frac{6.0 \times 10^{-12} \text{ g}}{\text{nucleus}} \times \frac{1 \text{ pair}}{1.16 \times 10^{-21} \text{ g}} = \mathbf{5.17 \times 10^9} \text{ **pairs per nucleus.**}$$

(d) $$\frac{5.17 \times 10^9 \text{ pairs}}{\text{nucleus}} \times \frac{1 \text{ turn}}{10 \text{ pairs}} \times \frac{3.4 \text{ nm}}{\text{turn}} = 17.6 \times 10^8 \text{ nm} = \mathbf{1.76} \text{ **meters of DNA.**}$$

3-10. (a)   Compared to a linear molecule, a branched-chain polymer has more termini for addition or hydrolysis of glucose units per unit volume of polymer, thereby facilitating both the deposition and mobilization of glucose by providing more sites for enzymatic activity.

(b)   Every branch point will have an $\alpha(1 \rightarrow 6)$ glycosidic bond that will have to be hydrolyzed. This is handled by the presence of an additional enzyme specific for the $\alpha(1 \rightarrow 6)$ bond.

(c)   Endolytic cleavage breaks the molecule infernally, creating additional ends for exolytic attack and thereby allowing the mobilization of more glucose per unit time.

(d)   Cellulose molecules are rigid, linear rods that aggregate laterally into microfibrils. Branches in the molecule would generate side chains which would almost certainly make it difficult to pack the cellulose molecules into microfibrils, thereby decreasing the rigidity and strength of the microfibrils.

3-11. See Figure S3-1 (p. 12) for the structures of (a) gentiobiose, (b) raffinose, and (c) a portion of a dextran chain.

3-12. (a)   Both maltose and lactose are reducing sugars because the glucose monomer on the right in Figure 3-23 in the textbook, when drawn in the straight-chain form, has a free aldehyde group on carbon 1 that can be oxidized to a carboxyl group by reaction with cupric ions. Sucrose, on the other hand, consists of a glucose monomer with its carbon atom 1 involved in the glycosidic bond and a fructose monomer that is a ketosugar rather than an aldosugar and therefore does not have a free aldehyde group in its straight-chain form.

(b)   Gentiobiose has a glucose monomer with a free carbon atom 1 and is therefore a reducing sugar. Raffinose is not a reducing sugar for the same reason cited in part (a) for fructose.

3-13. (a)   The two polymers have very different properties because they differ in structure as a result of differing $1 \rightarrow 4$ linkages between glucose monomers. The $\beta(1 \rightarrow 4)$ linkage of cellulose gives it a rigid, fibrous structure not seen in starch, which has an $\alpha(1 \rightarrow 4)$ linkage between glucose monomers.

(b)   The rigid, fibrous structure of cellulose and its virtual insolubility in water make this polymer a suitable component of the plant cell wall, while the more flexible structure of starch and its greater solubility in water make this polymer a suitable storage macromolecule.

(a)   Gentiobiose: (b) Raffinose:

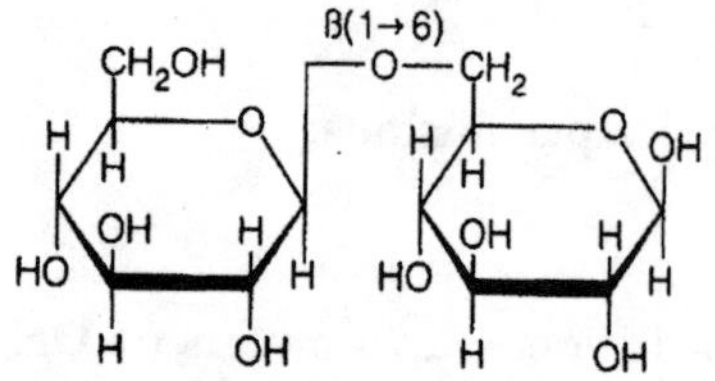

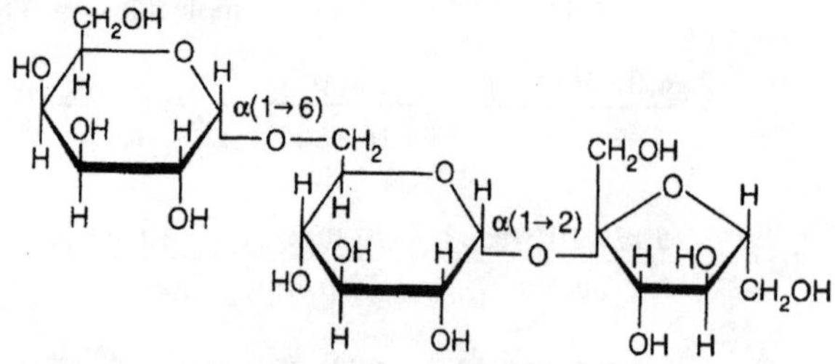

(c)   Portion of a dextran:

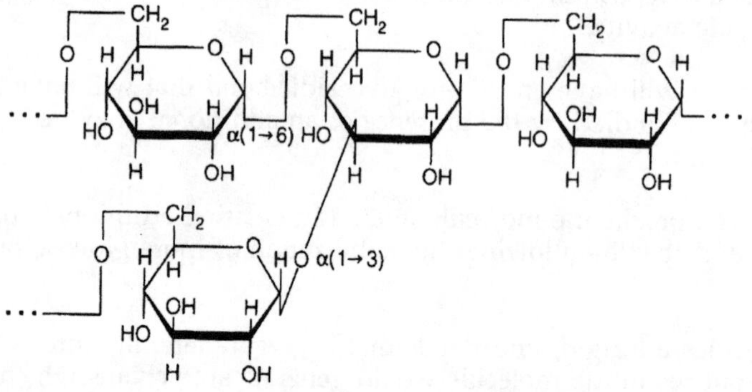

Figure S3-1 Carbohydrate Structures. The structures of the carbohydrates (a) gentiobiose, (b) raffinose, and (c) a portion of a dextran. See Problem 3-11.

3-14. (a)   The chemical formula for cellulose is written as shown because a molecule of water is removed every time a molecule of glucose is added to a growing cellulose polymer. Thus, $C_6H_{12}O_6$ becomes $C_6H_{10}O_5 + H_2O$.

(b)   The weight of the paper is 4 g. One mole (6.023×10^{23} molecules) of glucose units in the cellulose polymer weighs (72 + 10 + 80 =) 162 g, so 4 g. of glucose contain $(4.0/162) \times (6.023 \times 10^{23}) = \mathbf{0.15 \times 10^{23}}$ **glucose units.**

(c)   The length of each glucose unit is 0.45 nm or 0.45×10^{-9} meter and the length of the paper is 28 cm or 0.28 meter. The number of glucose units required is therefore $0.28/(0.45 \times 10^{-9}) = 0.62 \times 10^9 = \mathbf{6.2 \times 10^8}$ **glucose units.**

(d)   The width of each glucose unit is 0.30 nm or 0.30×10^{-9} meter and the width of the paper is 21.5 cm or 0.215 meter. The number of strands required is therefore $0.215/(0.30 \times 10^{-9}) = 0.72 \times 10^9 = \mathbf{7.2 \times 10^8}$ **glucose strands.**

3-15. (a)   A lipid is a molecule that is preferentially soluble in an organic solvent rather than in water. This definition is therefore based on the solubility properties of the molecule rather than on the chemical nature of the subunits or the bond that links subunits together, as is the case for proteins, nucleic acids, and carbohydrates.

(b)   Phosphatidyl choline > Fatty acid > Triacylglycerol > Estradiol > Cholesterol

(c)   The presence of an oleate will introduce a bend in one of the side chains that will closely approximate the shape of sphingomyelin, because the latter also has a double bond and hence a bend.

(d)   −11°C: Linolenic acid   +63°C: Palmitic acid

   +5°C: Linoleic acid   +70°C: Stearic acid

   +16°C: Oleic acid   +76.5°C: Arachidic acid

(e)   Phosphatidyl serine: the phosphoserine group

   Sphingomyelin: the phosphocholine group

   Cholesterol: the hydroxyl group (only slightly hydrophilic)

   Triacylglycerol: the ester bonds between glycerol and the fatty acids.

3-16.   (a)   Partial hydrogenation results in the partial reduction of C=C double bonds to C–C single bonds in the fatty acyl chains of the triglycerides.

(b)   Before partial hydrogenation, the shortening was vegetable oil and therefore liquid at room temperature.

(c)   Partial hydrogenation makes the shortening solid instead of liquid and allows it to be used as a substitute for animal fat.

4

Cells and Organelles

4-1.　(a)　T; the volume of a eukaryotic cell is usually several orders of magnitude greater than that of a prokaryotic cell.

(b)　T; some egg cells, neurons, and algal cells are examples.

(c)　F; a plasma membrane is common to all cells.

(d)　T; the ratio decreases with increasing cell size.

(e)　T; the ribosomes found in eukaryotic organelles are very similar to those present in prokaryotic cells.

(f)　F; these functions also occur in prokaryotic cells, but without compartmentalization into organelles.

4-2.　Sample calculations (for mycoplasma):

(i)　to span thumbtack: 1.2 cm/0.3 μm = $(1.2 \times 10^{-2})/(0.3 \times 10^{-6})$ = 4×10^4 = **40,000 cells.**

(ii)　to form layer: $\pi(0.6 \text{ cm})^2/\pi(0.15 \ \mu\text{m})^2 = (0.6 \times 10^{-2})^2/(0.15 \times 10^{-6})^2 =$ $(0.36 \times 10^{-4})/(0.0225 \times 10^{-12}) = (0.36/0.0225) \times 10^8 = 16 \times 10^8$ = **1.6 billion cells**

	Cell or Organelle	To Span Head of Thumbtack	To Cover Head of Thumbtack
		thousands	millions
(a)	Mycoplasma cell	40	1,600
(b)	Liver cell	0.6	0.36
(c)	Nucleus	2.0	4.0
(d)	Chloroplast	1.5	7.1
(e)	Mitochondrion	6.0	56.5
(f)	Peroxisome	24	576
(g)	Microtubule	12	4,520
(h)	Microfilament	60	81,000
(i)	Ribosome	400	160,000

4-3.   (a)   Secretion; pancreatic cells synthesize a variety of digestive enzymes and hormones, which are then secreted into the intestinal tract (enzymes) or the bloodstream (hormones).

(b)   Motility; muscle cells are capable of contraction, using the energy of ATP to cause movement.

(c)   Photosynthesis; palisade cells are the site of much photosynthetic activity in the leaf (see Problem 4-9).

(d)   Absorption; intestinal mucosal cells are especially well-suited for this function because they contain microvilli that greatly increase the absorptive surface area of the cell (see Figure 4-2 of the textbook).

(e)   Transmission of electrical impulses; the most important function of nerve cells is to conduct electrical signals from one part of the body to another.

(f)   Cell division; of the cell types listed, only bacterial cells are capable of rapid and repeated division, with only about 20-30 minutes between division under optimal conditions in at least some species.

4-4.   (a)   P   (d)   A, B, P   (g)   A, P   (j)   A, B, P

(b)   B, P   (e)   A, P   (h)   P   (k)   A, B, P

(c)   A, P   (f)   A, P   (i)   P   (1)   A, P

4-5.   (a)   Cellulose; chief component.

(b)   Prokaryotes; cyanobacteria are prokaryotes.

(c)   Hydrolases; lysosomal enzymes catalyze hydrolytic reactions.

(d)   Pores; the nuclear envelope is characterized by its pores.

(e)   Ribosomes; the nucleolus is the site of assembly of ribosomal subunits.

(f)   Secretory proteins; these proteins are synthesized by ribosomes bound to the rough ER.

g)   Lipid synthesis; occurs in smooth ER.

(h)   Glycoproteins; characteristic components of the plasma membrane.

(i)   Bacteriophage; viruses that infect bacterial cells.

(j)   RNA; the main component (and genetic material) of a viroid.

4-6.   (a)   look for plastids or a large vacuole.

(b)   become flaccid as water is drawn out of its cells.

(c)   a ribosome, microtubule, microfilament, etc.

(d)   salt water, hot springs, acidic environments, and sulfur-containing environments.

(e)   they are very similar in size.

4-7.   (a)   Look for the presence of a true nucleus and organelles such as endoplasmic reticulum, mitochondria, or lysosomes, all of which are present in a protist (eukaryote) but not in a bacterial cell (prokaryote).

(b)   Look for the presence of ribosomes, which are found on rough ER but not on smooth ER.

(c)   Analyze for the presence of any of the enzymes of photorespiration, which are present in leaf peroxisomes but not in animal peroxisomes.

(d)   Look for evidence of food particles inside the organelle; the presence of food particles is characteristic of lysosomes, but not late endosomes.

(e)   Look for the presence of a capsid (protein coat); a virus has an external protein layer, whereas a viroid does not.

(f)   Analyze for the presence of actin, which is the main protein of microfilaments but is not a component of intermediate filaments. (Alternatively, you might measure the diameter of the filament but that is not likely to be a definitive means of distinguishing between microfilaments and intermediate filaments because they do not differ very much in diameter.)

(g)   Analyze for the chemical nature of the nucleic acid (e.g., for the presence of ribose or deoxyribose); polio virus contains RNA as its genetic information, whereas herpes simplex virus is a DNA-containing virus.

(h)   Measure the sedimentation coefficient, which is about 70S for prokaryotic ribosomes and 80S for eukaryotic ribosomes.

(i)   Compare the sequences of ribosomal RNAs.

4-8.   (a)   B   (d)   N   (g)   A
        (b)   N   (e)   A   (h)   N
        (c)   B   (f)   N   (i)   B

4-9. (a)   Summary of calculations:

Structure	Dimensions	Volume	Number	Total Volume	Percent of Cell Volume
	μm	μm^3		μm^3	%
Cell	20×35	11,000	1	11,000	100.0
Vacuole	15×30	5300	1	5300	48.2
Nucleus	$d = 6$	113	1	113	1.0
Chloroplasts	2×8	25	40	1000	9.1
Mitochondria	1×2	1.570	200	314	2.9
Peroxisomes	$d = 0.5$	0.065	100	6.5	0.06
Ribosomes	$d = 0.025$	8.2×10^{-6}	2×10^6	16.3	0.14

(b)   Volume remaining = $100 - (48.2 + 1.0 + 9.1 + 2.9 + 0.06 + 0.14)$

$= 100 - 61.4 = \textbf{38.6\%}.$

This volume must accommodate the Golgi complex, the endoplasmic reticulum, the cytoskeleton, etc.

4-10. (a) 3   (c) 5   (e) 4   (g) 2

(b) 7   (d) 1   (f) 6

4-11. (a)   Mitochondrion; underactive due to "poisoning" of the electron transport system by cyanide (which binds to the terminal oxidase in the system, preventing transfer of electrons to oxygen).

(b)   Peroxisome; underactive due to the absence or inactivity of enzymes involved in the catabolism of fatty acids with very long hydrocarbon chains.

(c)   Nucleus; overactive in mitotic cell division.

(d)   Microtubule; underactive because of defective protein components, which prevent or reduce the motion of the flagella that are responsible for the motility of sperm cells; hence, the sperm cells' effectiveness in reaching and fertilizing an egg cell is compromised.

(e)   Lysosome; underactive due to the absence of the hydrolase enzyme and the consequent lack of ganglioside degradation.

(f)   Extracellular enzyme; underactive (or inactive) intestinal lactase (the enzyme that hydrolyzes lactose to glucose and galactose) means that lactose cannot be completely digested and absorbed in the small intestine and is converted by bacteria in the large intestine into toxic products that cause abdominal cramps and diarrhea. Eliminating milk from the diet usually alleviates these symptoms. (Lactose intolerance is common among adults of most human races except Northern Europeans and some Africans. It results from the disappearance after childhood of lactase from the intestinal cells.)

4-12.   (a)   Internal volume of a mycoplasma cell:

$$V = 4\pi r^3/3 = (1.33)(3.1416)(0.15)^3 = \textbf{0.014 } \mu\textbf{m}^3$$

Internal volume of an *E. coli* cell:

$$V = \pi r^2 h = (3.14)(0.5)^2(2.0) = \textbf{1.57 } \mu\textbf{m}^3$$

The bacterial cell has an internal volume that is more than a hundred times that of a mycoplasma cell $(1.57/0.014 = 112)$.

(b)   Assuming a prokaryotic ribosome to be roughly spherical with a diameter of about 25 nm (and therefore a radius of 12.5 nm or 0.0125 mm), its volume in cubic micrometers is:

$$V = 4\pi r^3/3 = (1.33)(3.14)(0.0125)^3 = \textbf{8.2} \times \textbf{10}^{-6} \ \mu\textbf{m}^3$$

If ribosomes represent 10% of the internal volume of a mycoplasma cell,the number of ribosomes that can be accommodated is:

$$(0.014 \ \mu m^3)(0.1)/(8.2 \times 10^{-6} \ \mu m^3) = 1.7 \times 10^2 = \textbf{170 ribosomes}$$

Not surprisingly, the total number of ribosomes present in a mycoplasma cell is much less than the number of ribosomes in a bacterial cell.

(c)   To use concentations in millimoles/liter, we need to convert the volume of the mycoplasm cell from cubic micrometers (mm^3) to cubic centimeters (cm^3) and then to L. We know that 1 $\mu m^3 = 10^{-6}$ meter and 1 cm $= 10^{-2}$ meter, so 1 $\mu m = 10^{-4}$ cm and 1 $\mu m^3 = 10^{-12} \ cm^3 = 10^{-12}$ mL. We can therefore estimate the internal volume of the mycoplasma cell to be $0.014 \times 10^{-12} \ cm^3$ or 0.014×10^{-12} mL or 0.014×10^{-15} L. At a concentration of 1.0 mM, the amount of glucose present in the cell is $(0.014 \times 10^{-15})(1.0 \times 10^{-3}) = 1.4 \times 10^{-20}$ moles, which represents $(6.023 \times 10^{23}$ molecules/mole$)(1.4 \times 10^{-20}$ moles$) = 8.43 \times 10^3 = \textbf{8430 molecules of glucose.}$

(d)   The concentration of NAD is only 0.002 $(2 \times 10^{-6}/1 \times 10^{-3})$ that of glucose, so the mycoplasma cell will contain only about $(8430)(0.002) = \textbf{17 molecules of NADH!}$

(e)   Weight of DNA molecule:

$(2 \times 10^7$ grams/mole$) \times 1$ mole$/(6.023 \times 10^{-23}$ molecules$) \times 1$ molecule $= 0.332 \times 10^{-16} = 3.32 \times 10\text{–}17$ gram.

Weight of mycoplasma cell: $(0.014 \times 10^{-12} \ cm^3)(1.1 \ g/cm3) = 1.54 \times 10^{-14}$ gram. The DNA therefore represents $(3.32 \times 10^{-17})/(1.54 \times 10^{-14}) = 2.16 \times 10^{-3}$ or **0.00216 (0.216%) of the mycoplasma cell by weight.**

The DNA consists of a circle with a circumference of about 10 mm or 10,000 nm and hence a diameter of about 3200 nm $(10,000/3.14)$, which is more than ten times the diameter of the mycoplasma cell!

(f)   The lower limit on cell size is probably determined by the need to have reasonable numbers of all the necessary molecules and to accommodate the amount of DNA required to be genetically autonomous.

4-13.   (a)   The nucleotide sequences of ribosomal RNAs—and hence of the genes that encode these RNAs—differ distinctively between eukaryotes and prokaryotes (and between eubacteria and archaebacteria as well) and are used as molecular criteria to distinguish between these fundamentally different kinds of organisms. Despite the large size of the E. fishelsoni cells, the nucleotide sequence of the genes that encode their ribosomal RNAs clearly identified this organism as a prokaryote.

(b)   A bacterium of this size is such a surprise because it was generally assumed that, in the absence of organelles to compartmentalize their function, prokaryotes cannot exceed a size limit set by the time it takes for molecules that need to react with each other to diffuse within the cytoplasm of the cell.

(c)   From Problem 4-12a, we know that the internal volume of an *Escherichia coli* cell is about 1.57 μm^3. The corresponding value for an *E. fishelsoni* cell can be calculated as:

$$V = \pi r^2 h = (3.14)(80/2)^2(600) = 3.0 \times 10^6 \ \mu m^3$$

The number of *E. coli* cells that could fit within the internal volume of one *E. fishelsoni* cell is therefore $(3 \times 10^6)/(1.57) = 1.9 \times 10^6$ = **1.9 million cells.**

CHAPTER

5

Bioenergetics:
The Flow of Energy
in the Cell

5-1.   (a) $(1.94 \text{ cal/min} \cdot \text{cm}^2)$ $(5.26 \times 10^5 \text{ min/year})$ $(1.28 \times 10^{18} \text{ cm}^2)$
$= \textbf{1.3} \times \textbf{10}^{\textbf{24}}$ **cal/year.**

      (b)   Some incident radiation is reflected back into space and much is absorbed by components of the earth's atmosphere. Atmospheric ozone plays an important role in filtering out ultraviolet radiation, whereas water vapor is responsible for most of the absorption in the infrared range.

      (c)   Much of the radiation falls on areas of the earth's surface where the climate is not favorable (too hot, too cold, too dry) for growth of phototrophic organisms during at least part of the year. In addition, about two-thirds of the earth's surface is covered by oceans which, though quantitatively significant in global photosynthesis, have in general only a very low density of phototrophic organisms and therefore a low efficiency of light utilization. Moreover, incident radiation represents a broad spectrum of wavelengths that can be used with varying degrees of efficiency by photosynthetic pigments, as discussed further in Chapter 15 of the textbook.

5-2.   (a)   Glucose is 40% carbon $(72/180 = 0.4)$, so 5×10^{16} g carbon represents about $\textbf{12.5} \times \textbf{10}^{\textbf{16}}$ **g organic matter.** (That is about 140 billion tons, if nonmetric units help you imagine the magnitude of the process.)

      (b) $(12.5 \times 10^{16} \text{ g})(3.8 \text{ kcal/g}) = 4.75 \times 10^{17} \text{ kcal} = \textbf{4.75} \times \textbf{10}^{\textbf{20}}$ **cal.**

      (c) $(4.75 \times 10^{20} \text{ cal})/(1.3 \times 10^{24} \text{ cal}) = 3.7 \times 10^{-4} = \textbf{0.037}\%$. (This means that more total solar energy will be received by the earth during the next 12 months than has been trapped photosynthetically since 700 BC!)

      (d)   Virtually all of it must be consumed by chemotrophs, because the earth is not undergoing any dramatic annual increase in amount of phototrophic organic matter accumulated.

5-3.   (a)   Use of blood sugar as a source of energy for muscle contraction; important for animal motility.

      (b)   Use of chemical energy to cause flash of light by firefly; important as mating signal.

(c)   Photosynthetic use of sunlight to synthesize sugar; entire biosphere depends on this process as its energy link with the sun.

(d)   Use of chemical energy to generate a potential and deliver an electric shock; defense mechanism of electric eel.

(e)   Use of chemical energy to pump protons into stomach and maintain low pH; aid to digestion of foodstuffs.

5-4.   (a)   The ΔH value of a reaction is a measure of the difference in heat content between the reactants and the products of the reaction and, therefore, of the heat exchange that accompanies the reaction under specified conditions. A negative ΔH value means that heat is liberated as the reaction proceeds. The ΔG value of a reaction is a measure of the difference in free energy between the reactants and products of the reaction and, therefore, of the exchange of free energy that accompanies the reaction under specified conditions. A negative ΔG value means that free energy is liberated as the reaction proceeds (and that the reaction is therefore thermodynamically spontaneous in the direction written).

(b)   The change in free energy, ΔG, is the algebraic sum of the change in enthalpy and the change in entropy that occur as reactants are converted into products under specified conditions. In other words, ΔG takes into account both the difference in heat content and the difference in entropy between the reactants and products, whereas ΔH measures only the former.

(c)   The reaction involves the conversion of a complex, 24-atom organic molecule into 12 simpler 3-atom molecules (with the additional 12 atoms provided by water). The distribution of carbon, hydrogen, and oxygen atoms is therefore much more random after the reaction than before. As a result, the entropy of the products is greater than that of the reactants, so the entropy change is *positive*.

(d) $\Delta G = \Delta H - T\,\Delta S$

$\Delta S = (\Delta H - \Delta G)/T$

$= (-673 - [-686])/(25 + 273) = +13/298 = +0.0436$

$= +43.6$ **cal/mol-K.**

The sign of ΔS agrees with the prediction in part c.

(e)   Because the overall reaction of photosynthesis is the reverse of reaction 5-26 in the textbook, the values of ΔS, ΔH, and ΔG will be identical in magnitude to the corresponding values for reaction 5-26, but opposite in sign.

5-5.   (a) $K'_{eq} = [F6P]/[G6P] = 0.5$

The concentrations of F6P and G6P after an overnight incubation in the presence of the enzyme catalyst will be:

$[F6P] = xM$ $[G6P] = (0.15 - x)M$

We can therefore write

$K'_{eq} = (x)/(0.15 - x) = 0.5$

Thus $1.5x = 0.075$ and $x = 0.075/1.5 = 0.05\ M$

Given that the F6P concentration is 0.05 mol/L or 0.05 mmol/mL,

10 mL of solution will contain **0.5 mmol of F6P.**

(b)   The same answer; the equilibrium concentrations of products and reactants depend only on the value of K'_{eq} and the temperature, not on the starting concentrations.

(c)   In the absence of a catalyst, the reaction would not proceed to any measurable extent, so the F6P concentration after the incubation would be negligible.

(d)   No, not without knowing the effect of temperature on the K'_{eq} value.

5-6.   (a) $\Delta G°' = -RT \ln K'_{eq} = (-1.987)(298) \ln (0.165) = -592(-1.802)$

$= +1067\ \textbf{cal/mol}.$

To convert 1 mole of 3-phosphoglycerate to 2-phosphoglycerate under conditions where the concentrations of both are maintained constant at 1.0 M would require the input of 1067 cal of free energy; the reaction is endergonic under standard conditions.

(b) $\Delta G' = \Delta G°' + (1.987)\ (298) \ln \dfrac{4.3 \times 10^{-6}}{61 \times 10^{-6}}$

$= +1067 + 592 \ln (0.0705) = 1067 + 592(-2.652)$

$= -500\ \textbf{cal/mol}.$

To convert 1 mole of 3-phosphoglycerate to 2-phosphoglycerate under the prevailing concentrations in the red blood cell will result in the liberation of 500 cal of free energy; the reaction is exergonic under prevailing conditions. $\Delta G'$ and $\Delta G°'$ differ because $\Delta G°'$ measures the energetics of the reaction under standard conditions whereas $\Delta G'$ measures the energetics of the reaction under actual, or prevailing conditions.

(c)   The concentration of 2-phosphoglycerate can rise only to the equilibrium value; if it goes above that, equilibrium will lie to the left, and the reaction will proceed in the reverse direction.

To calculate the equilibrium value:

$K'_{eq} = 0.165 = \dfrac{[2 - \text{phosphoglycerate}]}{61 \times 10^{-6}}$

$[2\text{-phosphoglycerate}] = (0.165)(61 \times 10^{-6)} = 10 \times 10^{-6}\ M$

$= \textbf{10}\ \mu\textbf{M}.$

(Alternatively, you can set $\Delta G'$ equal to zero and solve for [2-phosphoglycerate] using equation 5-16 in the textbook.)

5-7.   (a)   The $\Delta G°'$ value is positive, so the K_{eq}' value is less than one:

$\Delta G°' = (1.987)(298) \ln K_{eq}' = +1800$ cal/mol.

Thus, $\ln K_{eq}' = -1800/592 = -0.304$, so $K_{eq}' = 0.048$.

The equilibrium therefore lies to the **left**.

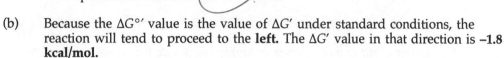

(b)   Because the $\Delta G°'$ value is the value of $\Delta G'$ under standard conditions, the reaction will tend to proceed to the **left**. The $\Delta G'$ value in that direction is **–1.8 kcal/mol**.

(c) $\Delta G' = \Delta G°' + RT \ln ([G3P]/[DHAP])$

$= +1800 + (1.987)(298) \ln (0.01) = +1800 + 592(-4.605) = 1800 - 2726$

$= -926$ cal/mol $= $ **–0.93 kcal/mol**.

(d)   Written in the direction required for the Calvin cycle, the reaction becomes

G3P $\rightleftharpoons$ DHAP $\Delta G°' = -1,800$ cal/mol

To calculate the ratio $X = [DHAP]/[G3P]$ at which the $\Delta G'$ will be $-3,000$ cal/mol in this direction, we can write

$-3,000 = -1,800 + RT \ln X$

So $592 \ln X = -3,000 + 1,800 = -1,200$ cal/mol

$\ln X = -1200/592 = -2.027$ and $X = $ **0.132**.

Thus, the [DHAP]:[G3P] ratio must be at or less than 0.132. If it is higher than this limit, the value of $\Delta G'$ will be less negative than $-3,000$ cal/mol. Conversely, the [G3P]:[DHAP] ratio must be at least $1/0.132 = $ **7.58**.

5-8.   (a) $\Delta G' = -RT \ln K_{eq}' = -7700$ cal/mol, so

$\ln K_{eq}' = (-7700)/(-592) = + 13.00$

$K_{eq}' = $ **4.42×10^5**

(Notice that K_{eq}' is a very large number, indicating that the equilibrium lies very far to the right, which is consistent with the highly negative value for $\Delta G°'$.)

(b) $K_{eq}' = [\text{citrate}]_{eq}[\text{coenzyme A}]_{eq}/[\text{OAA}]_{eq}[\text{acetyl CoA}]_{eq} = 4.42 \times 10^5$

$[\text{citrate}]_{eq}/3.0[\text{OAA}]_{eq} = 4.42 \times 10^5$

$[\text{citrate}]_{eq}/[\text{OAA}]_{eq} = (3)(4.42 \times 10^5) = 13.26 \times 10^5 = 0.1326 \times 10^7$

$[\text{OAA}]_{eq}/[\text{citrate}]_{eq} = 1/(0.1326 \times 10^7) = $ **7.54×10^{-7}**

(c) $\Delta G' = \Delta G^{\circ\prime} + RT \ln ([malate][coenzyme A]/[OAA][acetyl CoA]) = -2000$ cal/mol

$= -7700 + 592 \ln (O.4 \times 10^{-3})/(3)[OAA]) = -2000$

$= 592 \ln (0.133 \times 10^{-3})/[OAA]) = -2000 + 7700 = + 5700$ cal/mol

$\ln (0.133 \times 10^{-3})/[OAA]) = + 5700/592 = 9.628$

$(0.133 \times 10^{-3})/[OAA] = 1.519 \times 10^4$

$[OAA] = (0.133 \times 10^{-3})/(1.519 \times 10^{-4}) = 0.0876 \times 10^{-7} = 8.76 \times 10^{-9}$ M = **8.76 nM**

(Notice that this reaction is *so* highly exergonic that it will proceed to the right even when the malate concentration is less than 10 nM!)

(d) $K_{eq}{}' = [malate]_{eq}[coenzyme A]_{eq}/[OAA]_{eq}[acetyl CoA]_{eq} = 4.42 \times 10^5$

$= (0.4 \times 10^{-3})[coenzyme A]_{eq}/(8.76 \times 10^{-9})[acetyl CoA]_{eq} = 4.42 \times 10^5$

$[coenzyme A]_{eq}/[acetyl CoA]_{eq} = (4.42 \; 10^5)(8.76 \times 10^{-9})/(0.4 \times 10^{-3}) = 9.68$

$[acetyl CoA]_{eq}/[coenzyme A]_{eq} = 1/9.68$ = **0.103.**

The [acetyl CoA]/[CoA] ratio can drop to about 0.1 before the reaction will be at equilibrium and will therefore no longer proceed spontaneously to the right.

5-9.   (a)   Because $\Delta G^{\circ\prime} = 0$, the equilibrium constant must be 1.0. The reaction will therefore continue to the right until all species are present at equimolar concentrations of 0.005 M each.

(b)   Assume x mol/liter of succinate react with x mol/liter of FAD to generate x mol/liter each of fumarate and FADH$_2$.

At equilibrium:

[succinate] $= 0.01 - x$

[FAD] $= 0.01 - x$

[FADH$_2$] $= 0.01 + x$

[fumarate] $= x$

$$K'_{eq} = 1.0 = \frac{[fumarate] \; [FADH_2]}{[succinate] \; [FAD]} = \frac{(x)(0.01 + x)}{(0.01 - x)(0.01 - x)}$$

$$= \frac{0.01x + x^2}{0.0001 - 0.02x + x^2}$$

$0.01x + x^2 = 0.0001 - 0.02x + x^2$, so $0.03x = 0.0001$,

and therefore $x = \mathbf{0.0033}$.

Equilibrium concentrations are therefore:

[succinate] = 0.0067 M

[FAD] = 0.0067 M
[FADH$_2$] = 0.0133 M

[fumarate] = 0.0033 M.

(c) $\quad \Delta G' = \Delta G^{\circ\prime} + 592 \ln \dfrac{[\text{fumarate}]\,[\text{FADH}_2]}{[\text{succinate}]\,[\text{FAD}]} = -1500 \text{ cal/mol}$

$\quad = 0 + 592 \ln \dfrac{(2.5 \times 10^{-6})(5)}{[\text{succinate}]} = -1500 \text{ cal/mol}$

$\ln \dfrac{(12.5 \times 10^{-6})}{[\text{succinate}]} = \dfrac{1500}{592}$

so $\dfrac{(2.5 \times 10^{-6})}{[\text{succinate}]} = e^{-2.534} = 0.07934$

$12.5 \times 10^{-6} = 0.07934 \,[\text{succinate}]$

so [succinate] $= (12.5 \times 10^{-6})/0.07934 = 157 \times 10^{-6} \ M = \mathbf{157 \ \mu M}$.

5-10. (a) For the overall reaction A → D

$$K'_{AD} = \frac{[D]}{[A]} = \frac{[D]\,[B]\,[C]}{[A]\,[B]\,[C]} = \frac{[B]\,[C]\,[D]}{[A]\,[B]\,[C]} = K'_{AB} \cdot K'_{BC} \cdot K'_{CD}.$$

(b) $\Delta G^{\circ\prime}_{AD} = RT \ln K'_{AD} = -RT \ln [K'_{AB} \cdot K'_{BC} \cdot K'_{CD}]$

$\quad = -RT \ln K'_{AB} - RT \ln K'_{BC} - RT \ln K'_{CD} = \Delta G^{\circ\prime}_{AB} + \Delta G^{\circ\prime}_{BC} + \Delta G^{\circ\prime}_{CD}.$

(c) $\Delta G^{\circ\prime}_{AD} = \Delta G^{\circ\prime}_{AD} + RT \ln \dfrac{[D]}{[A]} = \Delta G^{\circ\prime}_{AD} + RT \dfrac{[D]\,[B]\,[C]}{[A]\,[B]\,[C]}$

$\quad = \Delta G^{\circ\prime}_{AD} + RT \ln \dfrac{[B]\,[C]\,[D]}{[A]\,[B]\,[C]}$

$\quad = \Delta G^{\circ\prime}_{AB} + \Delta G^{\circ\prime}_{BC} + \Delta G^{\circ\prime}_{CD} + RT \ln \dfrac{[B]}{[A]} + RT \ln \dfrac{[C]}{[B]} + RT \ln \dfrac{[D]}{[C]}$

$\quad = \Delta G^{\circ\prime}_{AB} + RT \ln \dfrac{[B]}{[A]} + \Delta G^{\circ\prime}_{BC} + RT \ln \dfrac{[C]}{[B]} + \Delta G^{\circ\prime}_{CD} + RT \ln \dfrac{[D]}{[C]}$

$\quad = \Delta G^{\circ\prime}_{AB} + \Delta G^{\circ\prime}_{BC} + \Delta G^{\circ\prime}_{CD}.$

5-11.   (a)   glucose + P_i $\rightarrow$ G6P + H_2O $\Delta G^{\circ\prime}$ = +3.3 kcal/mol

+ <u>(ATP + H_2O $\rightarrow$ ADP + P_i</u> $\Delta G^{\circ\prime}$ = –7.3 kcal/mol

glucose + ATP $\rightarrow$ G6P + ADP $\Delta G^{\circ\prime}$ = + 3.3 –7.3 = **–4.0 kcal/mol**

(b)   P-creatine + H_2O $\rightarrow$ creatine + P_i $\Delta G^{\circ\prime}$ = –10.3 kcal/mol

+ <u>(ADP + P_i $\rightarrow$ ATP + H_2O</u> $\Delta G^{\circ\prime}$ = +7.3 kcal/mol

P – creatine + ADP $\rightarrow$ creatine + ATP $\Delta G^{\circ\prime}$ = –10.3 + 7.3 = **–3.0 kcal/mol**

5-12.   (a) $\Delta G^{\circ\prime} = -RT \ln K'_{eq} = -592 \ln K'_{eq} = -4000$

$\ln K'_{eq} = -4000/(-592) = +6.757$

K'_{eq} = **860**

(b) $K'_{eq} = [G6P]_{eq}[ADP]_{eq}/[Glucose]_{eq}[ATP]_{eq} = 860$

$(0.05 \times 10^{-3})(0.15 \times 10^{-3})/[Glucose]_{eq}(2.0 \times 10^{-3}) = 860$

$(3.75 \times 10^{-6})/[Glucose]_{eq} = 860$

$[Glucose]_{eq} = (3.75 \times 10^{-6})/860 = 4.36 \times 10^{-9}$ = **4.36 nM.**

(c) $\Delta G' = \Delta G^{\circ\prime} + RT \ln [G6P][ADP]/[Glucose][ATP]$

$= -4000 + 592 \ln (0.05 \times 10^{-3})(0.15 \times 10^{-3})/(5.0 \times 10^{-3})(2.0 \times 10^{-3})$

$= -4000 + 592 \ln (7.5 \times 10^{-4}) = -4000 + (592)(-7.195)$

$= -4000 -4260 =$ **–8260 cal/mol.**

(d)   Let the $\Delta G^{\circ\prime}$ value for the direct phosphorylation of glucose by inorganic phosphate be X cal/mol. Then we can write two reactions which when summed give us reaction 5-39 in the textbook:

Glucose + P_i $\rightarrow$ Glucose-6-phosphate + H_2O $\Delta G^{\circ\prime}$ = X

<u>ATP + H_2O $\rightarrow$ ADP + P_i</u> $\Delta G^{\circ\prime}$ = –7300 cal/mol

Glucose + ATP $\rightarrow$ Glucose-6-phosphate + ADP $\Delta G^{\circ\prime}$ = –4000 cal/mol

Thermodynamic paraameters are additive, so we can write:

$X + (-7300) = -4000$

so $X = -4000 + 7300 = +3300$ cal/mol = **+3.3 kcal/mol**

This reaction is quite highly endergonic under standard conditions. It is unlikely that the intracellular concentrations of the reactants could be maintained high enough and/or the concentrations of the products maintained low enough to render the $\Delta G'$ negative. (To convince yourself of this, try making some reasonable assumptions about intracellular concentrations of the reactants and products and calculate the G' value under such conditions.)

(e)   A $\Delta G'$ value tells us nothing about the rate at which the reaction will proceed. It is a thermodynamic parameter and tells us simply how much energy will be liberated if the reaction proceeds. Whether the reaction will occur depends on the presence of the appropriate enzyme catalyst—hexokinase, in this case.

5-13. (a)   Because the folding process is thermodynamically spontaneous, the ΔG for protein folding is a **negative** value. The ΔG for the unfolding process is correspondingly **positive.**

(b)   Because the folding process converts a more random structure into a less random structure, the entropy will decrease and the ΔS for folding is therefore a **negative** value. Conversely, the ΔS for the unfolding process is a positive value.

(c)   Because the entropy contribution to ΔG is represented by the term $-T\Delta S$, a negative value of ΔS for the folding process will make the value of ΔG more positive. Conversely, a positive value for ΔS for the unfolding process will make the value of ΔG more **negative.**

6

Enzymes:
The Catalysts of Life

6-1. (a) It means that the molecules that ought to react would release energy if they were to do so, but do not possess enough energy to collide in a way that allows the reaction to be initiated.

(b) Touching a match to a sheet of paper is an example. Thermal activation imparts sufficient kinetic energy to the molecules such that the proportion of them that possess adequate energy to collide and react increases significantly. Once initiated, the reaction is self-sustaining, because reacting molecules release sufficient energy to energize and activate neighboring molecules for reaction.

(c) Interaction of molecules is facilitated (by positioning on the catalyst surface, for example), thereby requiring less energy to activate each molecule and so ensuring that substantially greater numbers of molecules possess adequate energy to initiate reaction without any elevation in temperature.

(d) Advantages: specificity, more exacting control.
Disadvantages: much more susceptible to inactivation by heat, extremes of pH, and so on; also, much energy needs to be expended to synthesize the enzyme molecules.

6-2. (a) The activation energy diagram for the catalase reaction is shown in Figure S6-1. To say that platinum lowers the activity energy from 18 to 13 kcal/mol means that platinum is a surface on which molecules of hydrogen peroxide are positioned such that their reaction to form water and molecular oxygen is more favorable than when hydrogen peroxide is free in solution. Specifically, it means that a pair of H_2O_2 molecules on a platinum surface requires only about 72% as much activation energy to initiate a reaction between them as when they are present in solution. As a result, many more of the H_2O_2 molecules possess adequate energy to interact at the prevailing temperature. Catalase provides a surface that is even more conducive to reaction; a pair of H_2O_2 molecules at the active site of catalase requires only about 39% as much activation energy as when the molecules are present in solution.

(b) The active site of the enzyme provides a surface that is very specific for the binding of H_2O_2 molecules. The particular amino acids and the iron-porphyrin complex that are present at the active site create a chemical environment with the right functional groups and pH to favor maximally the binding of H_2O_2 molecules and their subsequent reaction to form the desired products.

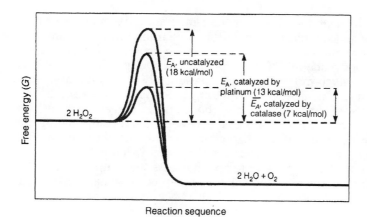

Figure S6-1   Effect of Catalysis on the Activation Energy. The activation energy E_A for the decomposition of H_2O_2 to H_2O and O_2 is about 18 kcal/mol. This value is reduced to about 13 kcal/mol by platinum, a metallic catalyst, and to about 7 kcal/mol by the enzyme catalase. See Problem 6-2a.

(c)   The rate of hydrogen peroxide decomposition can also be accelerated by an increase in temperature. However, this is not an appropriate means of increasing reaction rates within cells because most cells can function only within a rather limited temperature range; for most cells, increases in temperature lead to the denaturation of proteins and hence to cellular dysfunction or death.

(d)   The advantage of having H_2O_2-generating enzymes compartmentalized together with catalase within a membrane-enclosed organelle is that the hydrogen peroxide formed by these enzymes is generated within the immediate proximity of the enzyme that degrades it to water and oxygen. Moreover, the membrane around the peroxisome prevents the hydrogen peroxide from diffusing away and causing damage by oxidizing cellular constituents; instead, the H_2O_2 accumulates only within the peroxisomes, where an increase in its concentration will result in an immediate increase in catalase activity, as dictated by the Michaelis-Menten equation.

6-3.   (a)   To decompose the same quantity of H_2O_2 in the presence of an equivalent amount of ferric ions would take $10^8/(3 \times 10^4)$ or 3.33×10^3 times longer:

$3.33 \times 10^3 \times 1$ minute $\times 1$ hour$/60$ min = **55.5 hours.**

(b)   In the absence of a catalyst, the process would take 10^8 times longer:

$10^8 \times 1$ minute $\times 1$ hour$/60$ min $\times 1$ day$/24$ hour $\times 1$ year$/365.25$ days = **190 years.**

(c)   It should be clear that having to wait hours for a chemical reaction to occur to a significant extent is not compatible with the rapid changes that cells must be able to effect and that hundreds of years is in the realm of the ridiculous!

6-4.   (a)   Figure 6-5a in the textbook illustrates the temperature dependence of a typical human enzyme and a typical enzyme from a thermophilic bacterium. Each has a temperature optimum that is at or near the temperature of the organism (37°C for the human body, about 75°C for a typical hot spring). Clearly, the bacterial enzyme must be much more heat-stable, presumably due to the number and strength of both noncovalent bonds (hydrogen bonds and electrostatic interactions) and covalent (disulfide) bonds. In both cases, the reaction velocity increases steadily as the temperature is increased from a colder temperature to the temperature optimum of the enzyme, consistent with the effect of temperature on chemical reactions in general, which usually double in reaction velocity for every 10°C increase in temperature. As the temperature is raised above the optimum, however, the enzyme undergoes thermal denaturation resulting in loss of activity.

Figure 6-5b in the textbook illustrates the pH dependence of two enzymes with very different pH optima. The pH optimum for an enzyme corresponds to the proton concentration at which ionizable groups on the enzyme and/or the substrate molecules are in the most favorable form for chemical reactivity. Changes in pH away from the optimum result in loss of enzyme activity due to ditration of the ionizable groups on the enzyme and/or the substrate.

(b)   For Figure 6-5a, both enzymes are maximally active at or near the temperature of the milieu in which they are found—the human body in one case, a thermal hot spring in the other. For Figure 6-5b, the difference in pH optima for the two enzymes reflects the very different environments in which the two enzymes are active (the stomach for pepsin, the small intestine for trypsin).

(c)   An enzyme with a very flat pH profile probably has no amino acids at its active site that undergo ionization or protonation and it probably catalyzes a reaction in which neither the substrates nor the products can be ionized or protonated.

6-5.   (a)   A proteolytic enzyme is specific for a particular kind of bond *between* amino acids and is apparently little influenced by the exact chemical nature of the amino acid side groups surrounding that bond. A dehydrogenase attacks a specific bond *within* a molecule and apparently has an active site that is very specific for the particular molecule rather than the particular bond.

(b)   The subtilisin active site probably recognizes only elements of the peptide bond itself, whereas that of trypsin must have a binding site that is specific for a positively charged amino acid.

(c)   The structural analogues can bind to the active site, thereby rendering it unavailable for true substrate and as a result inoperative in catalysis. In effect, they render inoperative every active site they occupy for as long as they remain there.

6-6.   (a)   C   (c)   B   (e)   C
       (b)   A   (d)   B   (f)   A

6-7.   (a)   If product accumulation becomes significant, the back-reaction may begin to occur, and the net activity in the forward direction will be correspondingly reduced.

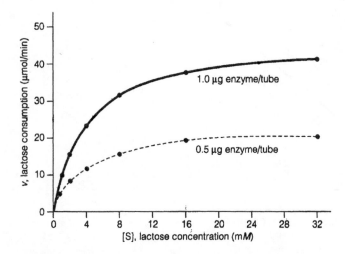

Figure S6-2   Kinetics of the β-Galactosidase Reaction. The dependence of initial reaction velocity on substrate concentration is shown for the data of Problem 6-7 with two concentrations of enzyme: 1.0 μg per tube (solid line), and 0.5 μg per tube (dashed line).

(b)   See the graph in Figure S6-2 (solid line). Doubling of substrate concentration always results in less than a twofold increase in velocity because the curve follows a hyperbolic equation, such that the relationship between substrate concentration and velocity is not linear at any point.

(c)   The data for the double-reciprocal plot are calculated as shown in Table S6-1. For the double-reciprocal plot, see the graph in Figure S6-3 (solid line).

Table S6-1. Double-Reciprocal Data for Problem 6-7.

$[S]$	$1/[S]$	v	$1/v$
mM	1 / mM	μmol / min	min/ μmol
1.0	1.000	10.0	0.100
2.0	0.500	16.7	0.060
4.0	0.250	25.0	0.040
8.0	0.125	33.3	0.030
16.0	0.0625	40.0	0.025
32.0	0.03125	44.4	0.022

(d) K_m: X-intercept is –0.25, so $K_m = -1/-0.25 = $ **+4.0 mM**.

V_{max}: Y-intercept is 0.02, so $V_{max} = 1/0.02 = $ **50 μmol/min**.

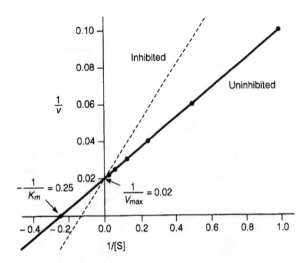

Figure S6-3   Double-Reciprocal Plot for the β-Galactosidase Reaction. Reciprocal values for the initial reaction velocity v and the substrate concentration [S] were determined for the data of Problem 6-7 and plotted as $1/v$ vs. $1/[S]$ (solid line). In the presence of a competitive inhibitor that increases the apparent K_m value by a factor of two, the line will be twice as steep (dashed line).

(e)   Results with 0.5 µg enzyme per tube are shown as the dashed line on the graph of Figure S6-2. Half as much enzyme results in half the total rate of lactose hydrolysis at any concentration of lactose (that is, reaction velocity is linear with enzyme concentration).

6-8.   (a)   In a double-reciprocal plot such as Figure 6-22 in the textbook, K_m is the negative reciprocal of the x-intercept: $K_m = -1/(20) = + \textbf{0.05 m\textit{M}}$.

This value tells us that the enzyme galactokinase will be functioning at 50% of its maximum velocity when galactose is present at a concentration of 0.05 mM.

(b) V_{max} corresponds to the reciprocal of the y-intercept:
$V_{max} = 1/0.1 = \textbf{10 µmol/min.}$

This value tells us that, for the specific amount of the enzyme used in the assay, as the substrate concentration tends toward infinity, the reaction velocity will tend toward (i.e., will asymptotically approach) 10 mmol/min.

(c)   You would expect to get the same V_{max} value because that is the velocity obtained when both substrates are present at high, nonlimiting concentrations, which will be true at the highest substrate concentrations used in both cases.

(d)   The K_m values for galactose and ATP are measures of the affinity of the enzyme for these two substrates. There is no *a priori* reason to assume that the enzyme would have the same affinity for two substrates, especially when they are chemically very different from each other.

6-9. (a) By dividing both sides of the Michaelis-Menten equation by V_{max}, we generate the following expression: $v/V_{max} = [S]/([S] + K_m)$. Given that K_m is 5 μM, the calculations are as follows:

   (i) $v/V_{max} = 0.5/(0.5 + 5.0) = 0.5/5.5 =$ **0.09 (9%)**

   (ii) $v/V_{max} = 5/(5 + 5) = 5/10 =$ **0.50 (50%)**

   (iii) $v/V_{max} = 50/(50 + 5) = 50/55 =$ **0.91 (91%)**

   Notice that the velocity of an enzyme-catalyzed reaction increases by an order of magnitude (from about 9% to about 90% of V_{max}) as the substrate concentration increases by two orders of magnitude around the K_m value. This is a general property of an enzyme-catalyzed reaction that follows Michaelis-Menten kinetics.

   (b) For tubes A and B, the Michaelis-Menten equation can be written as:

   Tube A: Tube B:

   $V_A = V_{max} [S]/([S] + K_m)$ $V_B = V_{max} [S]/([S] + K_m)$

   $\quad = V_{max} (0.1)/(0.1 + K_m)$ $\quad = V_{max} (0.2)/(0.2 + K_m)$

   Knowing that $V_B = 1.5 V_A$, we can write

   $V_B = V_{max} (0.2)/(0.2 + K_m) = 1.5 V_{max} (0.1)/(0.1 + K_m)$, or

   $(0.2)/(0.2 + K_m) = 1.5(0.1)/(0.1 + K_m)$

   Solving this expression algebraically for K_m yields

   $K_m =$ **0.2 mM.**

   (c) If initial reaction velocities were not specified, accumulation of products and/or depletion of substrate would decrease the reaction velocities and lead to an erroneous value for K_m.

   (d) To determine K_m for NAD^+, vary the NAD^+ concentration over a range of at least two orders of magnitude while holding the concentration of malate constant at some high, non-limiting value (about two orders of magnitude above its K_m value, if possible). (Alternatively, you could use just two NAD^+ concentrations and proceed as in part b, but bear in mind that the calculated value of K_{m, NAD^+} would then be based on only two data points rather than the range of data points used to construct a Michaelis-Menten plot.)

6-10. (a) Number of moles of enzyme present:

   $2 \mu g \times 10^{-6}$ g/$\mu g \times 1$ mole/30,000 g. $= (2 \times 10^{-6})/(3 \times 10^4) = 6.67 \times 10^{-11}$ moles.

   Rate of CO_2 consumption:

   $(6.67 \times 10^{-11}$ moles$) (1 \times 10^6)$/sec $= 6.67 \times 10^{-5}$ mole/sec $=$ **0.0667 mmoles/second.**

(b)   One mole of a gas occupies 22.4 liters at STP, so 0.0667 mmoles of CO_2 occupies

(0.0667 mmoles) × 22.4 mL/mmole = 1.5 mL.

The rate of CO_2 consumption is therefore **1.5 mL/second.**

6-11.   All five statements are true!

6-12.   (a)   The "burst of energy" that is needed for muscle contraction is provided by an increased rate of ATP synthesis by the muscle cells. Because the most common substrate for ATP synthesis in a muscle cell is glucose and glucose is stored in the cell as glycogen, the enzyme glycogen phosphorylase must be activated to break down the glycogen to glucose (see Figure 10-24b in the text). Thus, epinephrine and glucagon, when secreted into the bloodstream (from the adrenal gland and the pancreas, respectively), reach the mitochondria and trigger the activation of glycogen phosphorylase (by converting it from the *b* to the *a* form) and the activated enzyme then degrades stored glycogen, releasing glucose molecules that are catabolized to generate the needed ATP.

(b)   If any molecules of glycogen phosphorylase are in the *a* form in the presence of high concentrations of either ATP or glucose, it makes sense for the enzyme molecules to be inhibited allosterically by the ATP or the glucose, because further glycogen catabolism isn't necessary if either ATP is already present at an adequate concentration to power muscle contraction or an adequate supply of glucose is already available to generate the needed ATP.

(c)   Formation of the pancreatic enzyme carboxypeptidase by trypsin-activated cleavage of its inactive precursor makes good sense because trypsin is a gastric enzyme, so its appearance in the intestine is almost certainly the result of the movement of partially digested food from the stomach to the small intestine, which is the site of carboxypeptidase activity.

6-13.   The point of departure for the derivation is the simple enzyme-catalyzed reaction of equation 1 and the Michaelis-Menten model for its mechanism shown in equation 2:

$$S \xrightarrow{\text{enzyme}} P \tag{1}$$

$$E_f + S \underset{k_2}{\overset{k_1}{\rightleftharpoons}} ES \underset{k_4}{\overset{k_3}{\rightleftharpoons}} P + E_f \tag{2}$$

The velocity v can be expressed as the rate of disappearance of substrate or the rate of appearance of product:

$$v = -\frac{d[S]}{dt} = +\frac{d[P]}{dt} \tag{3}$$

To derive the dependence of v on [S], we begin by writing equations expressing the rates of change in concentrations of S, *ES*, and P:

$$\frac{d[S]}{dt} = -k_1 [E_f][S] + k_2 [ES] \tag{4}$$

$$\frac{d[P]}{dt} = -k_3[ES] - k_4[P][E_f] \tag{5}$$

$$\frac{d[ES]}{dt} = k_1 [E_f][S] - k_2[ES] - k_3[ES] + k_4[P][E_f] \tag{6}$$

Since we are confining ourselves to the initial stages of the reaction when [P] is essentially zero, equations 5 and 6 simplify to:

$$\frac{d[P]}{dt} = k_3[ES] \tag{7}$$

$$\frac{d[ES]}{dt} = k_1 [E_1][S] - k_2[Es] - k_3[ES] \tag{8}$$

$$= k_1 [E_f][S] - [k_2 + k_2][ES]$$

To proceed we must now assume the steady state at which the enzyme-substrate complex is being broken down at the same rate at which it is being formed, such that the net rate of change in [ES] *is* zero:

$$\frac{d[ES]}{dt} = k_1 [E_f][S] - [k_2 + k_3][ES] = 0 \tag{9}$$

This can be rewritten as

$$k_1 [E_f][S] = [k_2 + k_3][ES] \tag{10}$$

Clearly, the total amount of enzyme present, E_t, is simply the sum of the free form E_f plus the amount of complexed enzyme ES:

$$E_t = E_f + ES \tag{11}$$

Hence,
$$E_f = E_t - ES \tag{12}$$

which, when substituted into equation 10, yields

$$k_1 [E_t - ES][S] = [k_2 + k_3][ES] \tag{13}$$

or
$$k_1 [E_t][S] - k_1[ES][S] = [k_2 + k_3][ES] \tag{14}$$

Rearranging this yields

$$k_1 [E_t][S] = k_1[ES][S] = [k_2 + k_3][ES] \tag{15}$$

$$= (k_1[S] + k_2 + k_3)[ES] \tag{16}$$

which can be rewritten as

$$[ES] = \frac{k_1[E_t][S]}{k_1[S] + k_2 + k_3} \tag{17}$$

This is useful because equations 3 and 7 tell us that the velocity v is simply k_3 [ES]:

$$v = k_3 \,[\text{ES}] = \frac{k_1 k \,[\text{E}_t][\text{S}]}{k_1[\text{S}] + k_2 + k_3} \tag{18}$$

$$= \frac{k_1[\text{E}_t][\text{S}]}{[\text{S}] + k_2 + k_3 / k_1} \tag{19}$$

For a given concentration of enzyme, E_t is a constant, so we can define two kinetic constants as follows:

$$V_{\max} = k_3[\text{E}_t] \qquad K_m = (k_2 + k_3)/k_1$$

Rewritten in these terms, equation 19 becomes

$$v = \frac{V_{\max}[\text{S}]}{[\text{S}] + K_m}$$

6-14. (a)

Eadie-Hofstee Plot	**Hanes-Wolff Plot**
Cross-multiplication of the Michaelis-Menten expression yields	Cross-multiplication of the Michaelis-Menten expression yields
$vK_m + v[S] = V_{\max}[S]$.	$V_{\max}[S] = vK_m + v[S]$.
Dividing by $[S]$ gives	Dividing by $v \cdot V_{\max}$ gives
$vK_m/[S] + v = V_{\max}$, so	$[S]/v = K_m/V_{\max} + (1/V_{\max})[S]$.
$v = V_{\max} - K_m\,(v/[S])$.	

(b)

Plot v versus $v/[S]$.	Plot $[S]/v$ versus $[S]$.
Y-intercept $= V_{\max}$	X-intercept $= -K_m$
Slope $= -K_m$	Slope $= 1/V_{\max}$
X-intercept $= V_{\max}/K_m$	Y-intercept $= K_m/V_{\max}$

(c) A typical Hanes-Wolff plot is shown in Figure S6-4. The Hanes-Wolff plot is statistically the most satisfactory means of linearizing kinetic data because it has the data points (i.e., the [S] values) spaced out linearly along the X-axis, as in the original Michaelis-Menten plot. Both of the other plots use the reciprocal of [S] for the X-axis, thereby giving too much weight to values determined at low substrate concentrations—which are the data points that are most prone to error because of the small numbers involved.

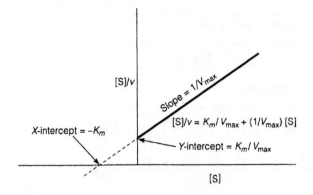

Figure S6-4 The Hanes-Wolff Plot. The ratio $[S]/v$ is plotted as a function of $[S]$. K_m can be determined from the X-intercept and V_{max} from the slope. See Problem 6-14.

7

Membranes: Their Structure, Function, and Chemistry

7-1. (a)   Localization of function   (f)   Regulation of transport

     (b)   Intercellular communication   (g)   Regulation of transport

     (c)   Regulation of transport   (h)   Detection of signals

     (d)   Localization of function   (i)   Intercellular communication

     (e)   Localization of function

7-2. (a)   Evidence of membrane asymmetry; 1950s.

     (b)   Evidence that the membrane permeability of solutes is related to how nonpolar they are; 1890s.

     (c)   Evidence that the phospholipid bilayer is not completely covered by surface layers of protein; 1960s.

     (d)   Evidence that such particles, when seen on biological membranes, are integral proteins; 1970s.

     (e)   Evidence that real membranes are not just phospholipid bilayers; 1920s.

     (f)   Evidence of two categories of membrane proteins differing significantly in location within the membrane and/or in their affinity for an aqueous environment; 1970s.

     (g)   Evidence that light-dependent proton pumping can be carried out by a relatively simple protein-pigment complex; in addition, the crystalline array allowed electron diffraction analysis, which led to the first understanding of how a membrane protein is organized within the lipid bilayer; 1970s.

7-3. (a)   If the lipids from 4.74×10^9 erythrocytes have an area of 0.89 m^2 (= 0.89×10^{12} μm^2). then each cell has a monolayer area of $0.89 \times 10^{12} / 4.74 \times 10^9$ = **188 μm^2**. This is almost twice the surface area of an erythrocyte as Gorter and Grendel estimated it at the time, leading to the conclusion that the surface of each cell was covered with two layers (i.e., a bilayer) of lipid.

     (b)   As it turned out, the lipid extraction technique that Gorter and Grendel used was not quantitative, so they underestimated the amount of lipid per cell. In fact, they extracted only about two-thirds of the total lipid from the erythrocytes, so both of their values were off by about the same extent—a classic case of the right conclusion but from flawed data.

7-4.   (a)   To calculate the number of protein molecules in the membrane, determine the number of moles and multiply by Avogadro's number:

$(6.0 \times 10^{-13}$ g$)$ $(1$ mole$/50{,}000$ g$)$ $(6.023 \times 10^{23}$ molecules$/$mole$)$

= **7.2 $\times 10^6$ protein molecules/membrane.**

(b)   Let p = number of grams of phospholipid in one membrane and c = number of grams of cholesterol in one membrane. Then $p + c = 5.2 \times 10^{-13}$ g.

Because phospholipid and cholesterol are assumed to be present in a 1:1 molar ratio, the number of moles of phospholipid ($p/800$) must equal the number of moles of cholesterol ($c/386$): $p/800 = c/386$. Solving these two simultaneous equations, we get:

$$p = 3.5 \times 10^{-13} \text{ g of phospholipid per membrane}$$

$$c = 1.7 \times 10^{-13} \text{ g of cholesterol per membrane.}$$

From these values we can calculate that a red cell membrane contains 2.6×10^8 molecules each of phospholipid and cholesterol, for a total of 5.2×10^8 lipid molecules per membrane. The lipid/protein ratio is therefore $(5.2 \times 10^8)/(7.2 \times 10^6)$ = **72 lipid molecules/protein molecule.** (This is a very representative value for the lipid/protein ratio of most plasma membranes.)

(c)   The phospholipid molecules occupy a total area of $(2.6 \times 10^8$ molecules$)$ $(0.55$ nm$^2/$molecule$)$ = 1.54×10^8 nm^2. Similarly, the cholesterol molecules occupy a total area of $(2.6 \times 10^8$ molecules$)$ $(0.38$ nm$^2/$molecule$)$ = 1.06×10^8 nm^2. Phospholipid and cholesterol therefore account for a surface area of 2.6×10^8 nm^2. The membrane is a bilayer, so this surface area is divided between two monolayers. The surface area per monolayer is therefore 1.3×10^8 nm^2. Because 1 nm^2 = 1×10^{-6} μm^2, the surface area occupied by lipid molecules can be expressed as 130 μm^2. This represents $130/145 \times 100\%$ = **90% of the total surface area of the membrane.** (The remaining 10% of the surface area is occupied by transmembrane proteins.)

7-5.   (a)   Palmitate: 16×0.13 nm = 2.08 nm

Laurate: 12×0.13 nm = 1.56 nm

Arachidate: 20×0.13 nm = 2.6 nm

(b)   The hydrophobic interior of a typical membrane is 4-5 nm.

Two molecules of palmitate laid end to end: 4.16 nm

Two molecules of laurate laid end to end: 3.12

Two molecules of arachidate laid end to end: 5.2 nm

Laurate (12C) molecules are too short to span the hydrophobic interior, whereas palmitate (16) and arachidate (20) are about the right length.

(c)   Each amino acid extends the long axis of the helix by about 0.56/3.6 = 0.156 nm. To span a length of 4.16 nm (two palmitate molecules) will require 4.16/.156 = 26.7 or **about 27 amino acids.**

(d)   Seven transmembrane segments of about 26.7 amino acids each account for about 187 amino acids, which represent 187/248 = 0.752, or **about 75% of the protein.** The remaining 61 (i.e., 248-187) amino acids are present in the six hydrophilic loops that link the seven transmembrane segments together, so the hydrophilic loops must contain an average of about 61/6 = **10 amino acids.**

7-6.   Only (b) and (e) are unlikely. (b) is unlikely because longer-chain fatty acids decrease membrane fluidity, which is the opposite effect needed when the temperature has been lowered. (e) is unlikely because bacteria do not contain cholesterol under any conditions. On the other hand, (a) is likely because membrane fluidity is temperature-dependent, and (c) and (d) are likely because unsaturated fatty acids will increase membrane fluidity.

7-7.   (a)   The molecular weight of the polypeptide is about $131 \times 110 = $ **14,400.**

(b)   The molecular weights of the carbohydrates are 180 for galactose, 180 for mannose, 179 for glucosamine, and 309 for sialic acid. A carbohydrate side chain containing 3 galactose units and 2 each of the other three molecules has a molecular weight of $(3 \times 180) + (2 \times 180) + (2 \times 179) + (2 \times 309) - (9 \times 18)$, where the latter term accounts for the 9 molecules of water that are removed when 9 molecules are condensed into a single chain and linked to the protein. Thus, each side chain has a molecular weight of $540 + 360 + 358 + 618 - 162 = 1714$, and the 16 such chains per molecule of glycophorin have an aggregate molecular weight of about **27,400.**

(c)   The total molecular weight of glycophorin is 14,400 + 27,400 = **41,800,** of which about **66%** (27,400/41,800 = 0.656) is contributed by the 16 carbohydrate side chains.

(d)   With two-thirds of its mass attributable to its carbohydrate side chains, glycophorin is indeed a "sweet" (the meaning of *glykos* in Greek) protein.

7-8.   Membrane 1 has uniformly long and saturated fatty acids, so it has the *highest* transition temperature of the three (41°C). Double bonds are very disruptive of phospholipid packing in the membrane, so membrane 2 has the *lowest* transition temperature (–36°C). The shorter fatty acid "tails" of membrane 3 will lower the transition temperature noticeably but not drastically, so this membrane will have the *intermediate* transition temperature (23°C).

7.9   (a)   Protein A probably has its 60 hydrophobic amino acids clustered into several hydrophobic stretches, each of which is likely to be a transmembrane segment. Protein B, on the other hand, almost certainly has most of its hydrophobic amino acids dispersed throughout the sequence, such that they become buried within the interior of the protein upon folding.

(b)   Proteins are much larger than phospholipids and therefore diffuse much more slowly within the membrane. Moreover, many membrane proteins are linked to other proteins, particularly to those in the cytosol just beneath the membrane, and are therefore not free to move regardless of how much time is allowed.

(c)   The cells near the hoof are likely to be exposed to colder temperatures and therefore require a higher proportion of unsaturated fatty acids in their membrane lipids to maintain membrane fluidity than do cells elsewhere in the reindeer leg.

(d)   Agents used as anesthetics are hydrophobic and can therefore intercalate into membranes, where their presence increases membrane fluidity. This makes the plasma membrane more permeable to ions, which then leak passively across the membrane, making it impossible to maintain the high sodium and potassium ion gradients necessary for transmission of nerve impulses.

7-10. (a)   Under these conditions, *Acholeplasma* cells can incorporate an appropriate combination of saturated and unsaturated fatty acids into their membranes to provide the cell with the optimum level of membrane fluidity.

(b)   Saturated fatty acids make a membrane less fluid. If only saturated fatty acids are available, the transition temperature of the membrane increases until the transition temperature is equal to the ambient temperature, at which point the membrane will gel.

(c)   The temperature of the culture could be raised to preserve membrane fluidity in spite of the prominence of saturated fatty acids in the membrane.

(d)   When a membrane gels, all cell functions that depend on the mobility of membrane proteins or lipids will be impaired or disrupted. Without the ability to transport solutes, detect and transmit signals, and carry out other membrane-dependent processes, a cell will not be able to survive.

(e)   Unsaturated fatty acids increase membrane fluidity, thus increasing the permeability of the membrane to ions and other solutes and making it impossible to maintain concentration gradients that are vital to life.

7-11. (a)   To determine the structure of a protein by X-ray crystallography requires that the protein be isolated and crystallized. It is relatively straightforward to isolate membrane proteins but obtaining them in crystalline form has proven difficult to do with most integral membrane proteins. However, if an integral membrane protein can be isolated and sequenced, hydropathy analysis can be used to infer structural information.

(b)   A hydrophobic amino acid residue such as valine or isoleucine has a positive hydropathy index, whereas a hydrophilic residue such as aspartic acid or arginine has a negative hydropathy index.

(c)   Isoleucine is the most hydrophobic of the four amino acids, so it has the highest positive value: 3.1. Arginine is the most hydrophilic, so it has the most highly negative value: –7.5. Because of its hydroxyl group, serine is slighty more hydrophilic than alanine and therefore has the more negative of the two remaining values: –1.1. Alanine is slightly hydrophobic and has a positive value: 1.0.

(d)   A horizontal bar should be drawn over each of the seven peaks in the hydropathy plot, each of which corresponds to a transmembrane segment. The average transmembrane segment is about 20-30 amino acids long, which compares favorably with the value of 27 calculated in Problem 7-5c. The protein is in fact bacteriorhodopsin; see Figure 7-21b in the textbook.

7-12.   (a)   Some of the membrane proteins are associated with the outer phospholipid layer of the plasma membrane and protrude out from the membrane sufficiently to allow exposed tyrosine groups to be labeled by the LP reaction.

(b)   Some of the membrane proteins associated with the outer phospholipid layer of the plasma membrane are glycoproteins, the carbohydrate side chains of which are accessible to the GO and borohydride.

(c)   All the glycoproteins of the erythrocyte membranes are associated with the outer phospholipid layer of the membrane, and at least a portion of every carbohydrate side chain protrudes from the membrane surface sufficiently far to be labeled.

(d)   All the proteins associated with the outer phospholipid layer of the plasma membrane are glycoproteins; proteins bearing no carbohydrate side chains are, without exception, inaccessible to the labeling reagents.

(e)   All major membrane proteins protrude at least to some extent on one side of the membrane or the other; none is totally buried in the interior of the membrane.

7-13.   (a)   You would expect no labeling, because we already know from Problem 7-12(c) that all glycoproteins are associated with the outer layer and would therefore be on the interior of an insideout vesicle.

(b)   You would expect to see labeling of all the proteins that were not labeled in Problem 7-12(a), because we know from Problem 7-12(e) that almost all membrane proteins are accessible from one side of the membrane or the other.

(c)   You would conclude that at least some proteins extend all the way through the membrane and actually protrude sufficiently on both sides of the membrane to allow them to be labeled on either side.

(d)   Label membrane proteins of intact cells with one of the techniques described, then prepare inside-out vesicles and use the second labeling technique.

8

Transport Across Membranes: Overcoming the Permeability Barrier

8-1. (a) F; the direction of movement depends on both the concentration gradient and the membrane potential, so facilitated diffusion can occur from a compartment of lower concentration to a compartment of higher concentration if the membrane potential is in the right direction and of sufficient magnitude.

(b) F; hydrolysis of high-energy phosphate bonds is a common means of driving active transport, but so is an ion gradient, as in sodium-driven cotransport of sugars and amino acids into cells.

(c) F; the K_{eq} for all uncharged solutes is 1.0; membrane permeability may affect the rate (or even the possibility) of movement but it has no effect on the concentration ratio if and when equilibrium is reached.

(d) T

(e) T

(f) T

(g) F; carbon dioxide and bicarbonate ions move in opposite directions across the erythrocyte membrane in most cases.

(h) F; if the sodium-potassium pump is inhibited, the sodium gradient necessary for sodium-driven glucose uptake cannot be maintained and cotransport will decrease or even cease as the sodium gradient collapses.

8-2. (a) 3, 7   (c) 2, 3, 4, 6, 7   (e) 1   (g) 8

(b) 2, 4, 6   (d) 4, 5   (f) 5

8-3. (a) A, F (c) D   (e) A   (g) N   (i) A, D, F

(b) D   (d) A   (f) D   (h) D, F   (j) A, F

8-4. Evidence that argues against the transverse carrier model: (a) Integral membrane proteins are embedded stably in the membrane and protrude from one or both sides based on their hydrophobic and hydrophilic regions; and (2) for a protein to traverse a membrane would require movement of its hydrophilic region(s) through the hydrophobic interior of the membrane, which would be highly endergonic and hence thermodynamically improbable.

8-5.   (a) $\Delta G_{inward} = RT \ln ([K^+]_{inside}/[K^+]_{outside}) = (1.987)(37 + 273) \ln 35$

$= 616 \ln (35) = (616)(3.55) = \textbf{+2190 cal/mol of potassium ions.}$

(b) $\Delta G_{inward} = RT \ln ([K^+]_{inside}/[K^+]_{outside}) + zFV_m$

$= +2187 \text{ cal/mol} + (1)(23{,}062)(-0.06 \text{ V})$

$= +2187 - 1384 = \textbf{+800 cal/mol of potassium ions.}$

(c)   Given the specified concentrations of ATP, ADP, and inorganic phosphate, we can calculate the free-energy change associated with the hydrolysis of one mole of ATP:

$\Delta G' = \Delta G°' + RT \ln ([ADP][Pi]/[ATP]) = -7300 + 616 \ln (0.01/5)$

$= -7300 + 616 \ln 0.002 = -7300 + (616)(-6.215) = -7300 - 3828$

$= \textbf{–11{,}100 cal/mol of ATP molecules.}$

Mathematically, the 11,100 cal of energy released by the hydrolysis is theoretically enough to drive the inward pumping of about 13.9 (11,100/800) moles of potassium ions. However, any pumping mechanism transports an integral number of ions per ATP molecule, so the maximum possible number is **13 potassium ions pumped per molecule of ATP hydrolyzed.** No known pumping mechanism actually achieves this ratio, however. Even the sodium/potassium pump that is responsible for most inward transport of potassium ions in animal cells moves only two potassium ions inward per mole of ATP hydrolyzed. (It is important to note, however, that this same pump moves sodium ions outward concomitantly, the energetics of which are considered in Problem 8-6c.)

8-6.   (a)   The sodium/potassium pump maintains gradients of both sodium and potassium ions using ATP as its energy source. Ion gradients can also be generated and maintained by exergonic electron transport, either during

(b)   An ion gradient is used to make ATP in both chemotrophic energy metabolism (Chapter 14) and photophosphorylation (Chapter 15).

(c)   Energy needed for outward pumping of sodium ions:

$\Delta G_{outward} = RT \ln ([Na^+]_{outside}/[Na^+]_{inside}) - zFV_m$

$= (1.987)(37 + 273) \ln (145/12) - (+1)(23{,}062)(-0.09 \text{ V})$

$= (616) \ln (12.1) - (-2076) = (616)(2.49) + 2076$

$= 1535 + 2076 = \textbf{3610 cal/mol of sodium ions.}$

Energy available from hydrolysis of ATP:

$$\Delta G' = \Delta G^{\circ\prime} + RT \ln ([ADP] [P_i]/[ATP]) = -7300 + 616 \ln (0.05/5)$$

$$= -7300 + 616 \ln 0.01 = 7300 + (616)(-4.605) = -7300 - 2837$$

= −10,100 cal/mol of ATP hydrolyzed.

Mathematically, the theoretical number of sodium ions that can be pumped outward per ATP hydrolyzed is 10,100/3610 = 2.80. However, any pumping mechanism transports an integral number of ions per ATP molecule, so ATP can be used to drive outward transport of Na^+ on a 2:1 basis but not on a 3:1 basis.

(d)   Energy needed to synthesize ATP:

$$\Delta G' = \Delta G^{\circ\prime} + RT \ln ([ATP]/[ADP] [P_i]) = 7300 + (1.987)(25 + 273) \ln (5/0.05)$$

$$= 7300 + 592 \ln 100 = 7300 + (592)(4.605) = 7300 + 2726$$

= 10,030 cal/mol of ATP synthesized.

Energy available from proton gradient:

$$\Delta G_{inward} = RT \ln ([H^+]_{inside}/[H^+]_{outside}) + zFV_m$$

$$= (1.987)(25 + 273) \ln (10^{-8}/10^{-7}) - (+1)(23,062)(-0.18 \text{ V})$$

$$= (592) \ln (0.1) - (-23062)(0.18) = (592)(-2.303) + 4151$$

$$= -1363 - 4151 = \textbf{−5514 cal/mol of protons.}$$

The energy available from the proton gradient is not adequate to support ATP on a 1:1 basis. It is, however, adequate to support ATP synthesis on a 1:2 basis because the energy yield of two moles of protons (2 × −5514 = −11,028 cal) is sufficiently exergonic (though only barely so) to provide the −10,030 cal needed for the synthesis of one mole of ATP.

8-7.   (a)   For ATP hydrolysis:

$$\Delta G = -7300 + RT \ln \frac{[ADP] [P_i]}{[ATP]}$$

$$= -7300 + (1.987) (298) \ln (0.002)(0.001)/(0.020)$$

$$= -7300 + 592 \ln (0.0001) = -7300 + (592)(-9.210)$$

$$= -7300 + (-5452) = -12,800 \text{ cal/mol ATP.}$$

Available per sodium ion:

$$\Delta G = (-12,750 \text{ cal/mol ATP})(1 \text{ mol ATP}/3 \text{ mol Na}^+)$$

$$= -4250 \text{ cal/mol Na}^+$$

For outward sodium movement:

$$\Delta G_{outward} = RT \ln \frac{[Na^+]_{outside}}{[Na^+]_{inside}} - zEF$$

$$= 592 \ln \frac{0.15}{[Na^+]_{inside}} - (+1)(-0.075)(23,062)$$

$$= 592 \ln \frac{0.15}{[Na^+]_{inside}} + 1730 = +4250 \text{ cal/mol.}$$

$$\ln \frac{0.15}{[Na^+]_{inside}} = (4250 - 1730)/592 = 2520/592 = 4.26$$

$$\frac{0.15}{[Na^+]_{inside}} = e^{4.26} = 70.8$$

$$[Na^+]_{inside} = 0.15/70.8 = 0.00212 = \textbf{2.12 m}\boldsymbol{M}.$$

(b)   If it were an uncharged molecule, the internal concentration could be reduced much more because the ATP-driven pumping would not have to "fight" the membrane potential. (The actual value for an uncharged molecule would be 0.11 mM.)

8-8.   (a)   The concentration gradient is $10^{-2}/10^{-7} = \textbf{10}^{\textbf{5}}$.

(b) $\Delta G_{outward} = RT \ln \dfrac{[H^+]_{inside}}{[H^+]_{outside}}$

$$= (1.987)(310) \ln (10^5) - (+1)(-0.07)(23,062)$$

$$= +7092 \text{ cal/mol} + 1613 \text{ cal/mol}$$

$$= +8705 \text{ cal/mol} = \textbf{+8.7 kcal/mol.}$$

(c) $\Delta G°$ for ATP hydrolysis is only -7.3 kcal/mol, but concentrations of ATP, ADP, and Pi are usually such that ΔG for ATP hydrolysis in the cell is in the range of -10 to-14 kcal/mol, which would probably be adequate to drive hydrogen secretion on a 1:1 basis.

(d)   If the membrane potential (E) is to be just high enough to prevent the inward movement of protons when the outside-to-inside proton gradient is 10^5, then we can write:

$$\Delta G_{inward} = RT \ln \frac{[H^+]_{inside}}{[H^+]_{outside}} + zEF = 0$$

or $E = (-RT/zF) \ln (10^{-5})$

$$= -(1.987)(310)/(+1)(23,062) \times \ln (10^{-5})$$

$$= (-0.0267)(-11.5) = +0.308 \text{ V} = \textbf{+308 mV.}$$

8-9   (a) $\Delta G_{inward} = RT \ln \frac{[Na^+]_{inside}}{[Na^+]_{outside}} + zEF$

$$= (1.987) (310) \ln (7.5/105) + (+1)(-0.065)(23,062)$$

$$= 616 \ln (0.0714) -1498 = -1626 - 1498$$

$$= -3124 \text{ cal/mol} = \textbf{-3.12 kcal/mol.}$$

(b) $\Delta G_{inward} = RT \ln \frac{[glycine]_{inside}}{[glycine]_{outside}} + zEF$

$$= (1.987) (310) \ln (15/0.1) + 0 \text{ (no net charge)}$$

$$= 616 \ln (150) = +3086 \text{ cal/mol}$$

$$= \textbf{+3.09 kcal/mol.}$$

Inward glycine transport requires the concomitant inward transport of at least one sodium ion.

(c) $\Delta G_{inward} = RT \ln \frac{[aspartate]_{inside}}{[aspertate]_{outside}} + zEF$

$$= (1.987) (310) \ln (22.5/0.15) + (-1)(-0.065)(23,062)$$

$$= 616 \ln (150) + 1498 = +3086 + 1498$$

$$= +4584 \text{ cal/mol} = \textbf{+4.58 kcal/mol.}$$

Inward aspartate transport requires the concomitant inward transport of at least 2 sodium ions.

(d)   The inward transport of aspartate requires more energy because, unlike glycine, aspartate is negatively charged at pH 7, so that both the aspartate concentration gradient and the membrane potential oppose uptake of this amino acid.

(e)   Neither aspartate nor glycine is likely to diffuse freely back through the epithelial cell membrane because both are highly polar molecules.

8-10.   (a)   Figure 8-16 shows that 0.2 μmoles of ATP were hydrolyzed during an interval of 1 minute (beginning at 2 minutes and ending at 3 minutes) by 1.0 mg of protein present as added sarcoplasmic reticulum. The ATPase activity is therefore **0.2 μmoles/min per milligram of protein.**

(b)   During the same 1-minute interval, all of the added calcium (0.4 μmoles) was taken up, as shown by the change in slope of the ATP hydrolysis rate at the end of the minute. The Ca^{2+}/ATP ratio is therefore 0.4 μmoles/0.2 μmoles = **2 calcium ions transported inward for each molecule of ATP that is hydrolyzed.**

(c)   When the ionophore is added at 4 minutes, the calcium that had been accumulated within the vesicles leaks out. The presence of calcium ions in the medium activates the calcium-dependent ATPase, so that ATP hydrolysis begins once again. The reaction will continue until all of the ATP has been hydrolyzed to ADP and phosphate.

8-11   (a)   Such vesicles are free of endogenous energy sources and do not metabolize most substrates. However, you cannot be sure that a given transport system will be active in such vesicles, and it is possible that membrane proteins could alter their positions during vesicle formation.

(b) Na^+: outside; K^+: inside; ATP: outside.

(c)   ATP hydrolysis will continue at a high constant rate until the sodium and potassium concentration gradients across the membrane approach the maximum levels attainable with the given concentration of ATP, ADP, and P_i. The rate will then drop rapidly to the low baseline level of whatever ATP hydrolysis is necessary to replenish the gradient due to ion leakage across the membrane.

8-12.   (a)   No; sodium ion cotransport is required for active transport of glucose (by the Na^+/glucose symporter), but not for facilitated diffusion of glucose (by the glucose transporter).

(b)   Yes; cotransport of sodium ions drives the inward movement of amino acids and can only occur if sodium ions are actively pumped back out again.

(c)   Yes; potassium ions must be actively pumped into red blood cells, and this can only occur via a pump that couples the inward pumping of potassium ions to the outward pumping of sodium ions.

(d)   No; active uptake of sugars and amino acids in bacteria is driven by a proton gradient and is therefore not coupled to sodium cotransport as it is in animal cells.

8-13.   The 12 hydrophobic segments presumably correspond to 12 transmembrane segments. Assuming that it takes about 25 amino acids to span a membrane as a helix, the transmembrane segments account for about 300 amino acids, or about 60% of the total length of the protein. The remaining 192 amino acids make up the 11 hydrophilic loops that connect the 12 transmembrane segments, as well as the C-terminal and N-terminal ends of the protein; these 13 hydrophilic segments apparently consist of about 15 amino acids each, on average. Transmembrane segments 3, 5, 7, 8, and 11 probably associate side-by-side, thereby forming a transmembrane channel, with the hydrophilic side of the helix oriented toward the center of the channel in each case. The remaining seven transmembrane segments presumably surround the five channel-forming segments, most likely stabilizing their organization and thereby stabilizing the membrane channel that they define.

9

Signal Transduction Mechanisms: I. Electrical Signals in Nerve Cells

9-1. (a) S   (c) S   (e) N   (g) N

      (b) A   (d) N   (f) S

9-2. (a) These are the only three ions to which the plasma membrane of the nerve cell is sufficiently permeable to warrant their inclusion in the equation.

      (b) A more general expression for monovalent ions is:

$$V_m = \frac{RT}{F} \ln \frac{\sum P_{cation}[cation]_{outside} + \sum P_{anion}[anion]_{inside}}{\sum P_{cation}[cation]_{inside} + \sum P_{anion}[anion]_{outside}}$$

      (c) With the relative permeability for sodium ions at 0.01, the value for V_m is –77 mV, calculated according to equation 9-5 in the textbook. If the relative permeability for sodium ions were 1.0 instead, the value for V_m would be about –6.9 mV.

      (d) No, because V_m is proportional to the logarithm of the expression that contains the term for sodium permeability.

9-3. (a) (5 pA) $(1 \times 10^{-12}$ ampere /pA) $(6.2 \times 10^{18}$ charges /amp-sec) (0.005 sec) = **155,000 charges.** Thus, 155,000 ions pass through a single channel during the 5 milliseconds that it is open.

      (b) No. Even though 155,000 ions pass through a single open channel, this is insufficient current to cause any perceptible change in the resting potential of a postsynaptic membrane. (An actual synaptic transmission event involves the simultaneous fusion of several hundred synaptic vesicles with the presynaptic membrane. Each vesicle releases several thousand neurotransmitter molecules, and each of these, in turn, causes thousands of receptor channels in the post-synaptic membrane to open for a few milliseconds. Thus, several hundred thousand channels open at once, generating a current that is sufficient to drive the resting potential of the postsynaptic membrane to its threshold, leading to depolarization of the membrane and an ensuing action potential.)

9-4. (a) E_{Cl} will be negative, because only a negative charge on the inner surface of the membrane would counteract the tendency of the concentration gradient to drive chloride ions inward.

(b) $E_{Cl} = (RT/zF) \ln \dfrac{[Cl^-]_{outside}}{[Cl^-]_{inside}} = -58 \log_{10} \dfrac{560}{50} = \mathbf{-60\ mV.}$

(Note that the RT/zF value is negative because $z = -1$ for an anion.)

(c)   If the internal chloride ion concentration is 150 mM instead of 50 mM, the value of E_{Cl} becomes less negative (-33 mV instead of -60 mV)

(d)   The chloride ion concentration is probably so variable because the positive charge of sodium and potassium ions can be balanced in part by other anions, notably proteins. To the extent that the protein content of the axon varies, so might the internal chloride concentration.

9-5.   (a) $E_K = \dfrac{(2.303)(1.987)(273 + 37)}{(+1)(23{,}062)} \log_{10} \dfrac{4.6}{150}$

$= +0.06157 \log_{10}(0.0306)$

$= -0.0932\ V = \mathbf{-93.2\ mV.}$

$E_{Na} = +0.06157 \log_{10} \dfrac{145}{10} = +0.0715\ V = \mathbf{+71.5\ mV.}$

$E_{Ca} = (+0.06157/2) \log_{10} \dfrac{6}{0.001}$

$= +0.0308 \log_{10}(6000)$

$= +0.116\ V = \mathbf{+116\ mV.}$

(b)   The resting potential of a cardiac muscle cell is substantially more negative than the squid axon because of a greater potassium ion concentration gradient across the membrane of the cardiac cell.

(c)   Either Na^+ or Ca^{2+}, or both; both would move *inward*.

(d)   You could remove either sodium or calcium ions from the surrounding medium and observe the effect this has on the action potential. Radioactive isotopes could also be used to follow specific ions.

(e)   Potassium is driven outward both by its concentration gradient across the membrane and by the temporarily positive membrane potential. At A, the voltage-dependent potassium gates are not yet open.

(f)   This should decrease the concentration gradient for potassium and therefore decrease the rate of potassium ion movement out of the cell. The resting potential for the cell should be less negative.

9-6.   (a)   Of the sodium channels in the membrane, at least some will open in response to a stimulus that depolarizes the membrane by about 20 mV.

(b) Depolarizations of less than 20 mV are not sufficient to start the positive feedback cycle that leads to an action potential, and they therefore elicit no response.

(c) Intensity of stimulus is detected as the frequency with which individual neurons respond and/or as the difference in the number of separate neurons that respond.

9-7. Saltatory propagation would be disrupted or slowed at the very least, since the insulating layer of myelin would be reduced or absent. In addition, voltage-gated channels are clustered at nodes of Ranvier in myelinated neurons. Loss of saltatory propagation might result in the inability to propagate action potential as far as the next node in the absence of insulating myelin.

9-8. An action potential cannot move backwards in the direction of membrane that has just experienced an action potential. Once a region of the membrane has experienced an action potential, it becomes temporarily unresponsive and incapable of undergoing another action potential. This is known as the refractory period. The refractory period can be divided into two periods known as the absolute and relative refractory periods. The absolute refractory period is due to the inactivation of sodium channels. Following the absolute refractory period is the relative refractory period where a new action potential is possible but difficult to initiate. The relative refractory period is due to the increased permeability to potassium ions which follows the opening of sodium channels during an action potential.

9-9. Increasing the permeability of the membrane to potassium causes the membrane potential to go more negative and thus makes it more difficult to depolarize the membrane to threshold. Opening chloride channels can inhibit excitability in two ways. First, increased chloride flux could cause the membrane potential to become more negative. Second, when chloride enters along with a sodium ion, there is no depolarization, so the faster chloride can enter, the more difficult it is to depolarize the membrane. Increasing the permeability to either potassium or chloride ions would therefore inhibit excitability.

9-10. (a) Competes with acetylcholine for binding to the acetylcholine receptor, thereby blocking depolarization of the postsynaptic membrane. Leaves membrane fully polarized.

(b) Forms a covalent complex with the active site of acetylcholinesterase, thereby preventing hydrolysis of acetylcholine. Leaves membrane depolarized.

(c) Mimics acetylcholine in binding to the acetylcholine receptor and causing depolarization of the postsynaptic membrane, but is hydrolyzed much more slowly. Leaves membrane depolarized.

(d) Inhibits acetylcholinesterase by forming a stable carbamoyl-enzyme complex at the active site, thereby preventing the hydrolysis of acetylcholine. Leaves membrane depolarized.

9-11. If an excitatory neurotransmitter remains in the synaptic cleft for prolonged periods of time, the postsynaptic neuron will experience an excitatory postsynaptic potential (EPSP) for a longer period of time than normal. This will result in a prolonged period during which a rapid succession of action potential will be generated in the postsynaptic neuron.

CHAPTER

10

Signal Transduction Mechanisms: II. Messengers and Receptors

10-1. (a)   Calmodulin

(b)   pituitary

(c)   ligand

(d)   inositol trisphosphate (InsP3) and diacylglycerol (DAG)

(e)   adenylyl cyclase; phosphodiesterase

10-2.   The activation of Ras requires the assistance of another protein called SOS. Upon binding to the tyrosine kinase receptor, SOS catalyzes the Ras GDP/GTP exchange reaction. In this respect, SOS and Ras together are similar to a heterotrimeric G protein. Here, SOS acts like the GBγ subunits, whereas Ras is similar to a Gα subunit.

10-3. (a)   When sodium rushes into a cell, the cell's plasma membrane experiences a depolarization. The change in electrical potential of the membrane results directly in a change in cell surface proteins, such as voltage-gated channels. In contrast, when cytosolic calcium levels rise, calcium ions or calcium-calmodulin complexes in the cytosol bind to target proteins within the cell, altering their function.

(b)   Most cells are able to maintain a very low Ca^{2+} concentration (in the range of 10^{-8} to 10^{-7} M) by using a calcium ATPase to move calcium ions outward. Because the cytosolic calcium concentration is so low, it can be changed dramatically and rapidly by admitting very small amounts of calcium and the concentration change can very effectively regulate specific cellular functions. For example, a ten-fold change in calcium concentration in a muscle cell can be achieved within milliseconds by releasing micromolar amounts of calcium ions from the sarcoplasmic reticulum into the cytoplasm. To achieve a ten-fold change in the concentration of a more abundant ion such as potassium or sodium would require the release of millimolar amounts of potassium ions within milliseconds, which is impossible.

10-4.   The chelator could be added to the extracellular medium to reduce the calcium outside the cell. Addition of the chelator to the extracellular medium should block the action of the hormone if it depends on calcium influx. If the action of the hormone depends solely on calcium release from intracellular stores such as the endoplasmic reticulum, then addition of a calcium ionophore should mimic the action of the hormone.

Furthermore, the ionophore should be effective even when EGTA is added to the extracellular medium to reduce the free calcium ion concentration.

10-5.   (a)   Prostaglandins are derived from arachidonic acid, which is one of the fatty acids normally present in membrane phospholipids. Both aspirin and cortisone can block the formation of prostaglandins. Aspirin blocks the enzyme cyclooxygenase that catalyzes the first step in the pathway whereby arachidonic acid is converted to prostaglandins. Cortisone causes a similar effect by inhibiting phospholipase A, the enzyme that releases arachidonic acid from membrane phospholipids.

    (b)   Anything that stimulates release of prostaglandins. This might include stimulators of phospholipase A or cyclooxygenase.

10-6.   The binding of epinephrine to β-adrenergic receptors causes a stimulation of heart function, both in terms of heart rate and with respect to the amount of work done in pumping blood. The effect appears to be mediated by cyclic AMP. When an antagonist such as the beta blocker propranolol is given to patients with hypertension, the cellular response caused by the binding of epinephrine to beta receptors is partially inhibited. Heart function is gradually restored over a period of time, with a corresponding decrease in blood pressure.

10-7.   Excess wild-type receptor is injected to make sure that the effects are not due to the non-specific effects of simple excess of any receptor protein. Excess wild-type receptor would be expected to cancel the effects of the mutant receptors because as the concentration of the normal receptors increases within the plasma membrane, the likelihood of two wild-type receptors dimerizing increases. This in turn lessens the dominant-negative effects of the mutant receptor protein.

10-8.   Pertussis toxin ADP ribosylates certain G proteins. It would be likely that a neutrophil responds to bacterial proteins through a G protein-linked receptor.

10-9.   The response to smoking is complex and involves a large number of changes in the function of neurons. However, smokers do develop tolerance to nicotine. Therefore, it would be reasonable to suggest that chronic exposure to nicotine might lead to receptor down-regulation as a basis for this tolerance.

10-10.   (a)   Activated phospholipase C_β would stimulate production of $InsP_3$ via the G protein-linked receptor pathway. Ca^{2+} release from the endoplasmic reticulum (ER) would be stimulated, and the egg would be activated in the absence of sperm.

    (b)   If these receptors were made in sufficient quantities, they would "compete" with normal receptors for available $InsP_3$ (i.e., they would act in dominant-negative fashion). The result would be reduced calcium release from the ER following fertilization.

    (c)   A wave of aequorin fluorescence would be observed to sweep across the egg, initiating at the site of sperm entry into the egg.

10-11.   (a)   If Bcl-2 mRNA (and hence protein) is no longer made, then there will be loss of Bcl-2 protein. Since this is a mitochondrial trigger for apoptosis, we would expect apoptosis to be inhibited.

(b)   TRAIL would activate pathways similar to TNF-α, a known stimulator of apoptosis.

(c)   Caspase-3 is a key activator of the apoptosis pathway, so inhibiting it would suppress cell death.

11

Beyond the Cell: Extracellular Structures, Cell Adhesion, and Cell Junctions

11-1. (a) Both the ECM of animal cells and the walls around plant cells consist of long, rigid fibers embedded in an amorphous, hydrated matrix of branched-chain molecules, either glycoproteins (ECM) or polysaccharides (cell wall).

(b) For ECM, the fibers consist of collagen and the matrix is a network of proteoglycans. For cell walls, the fibers consist of cellulose and the matrix is a network of polysaccharides and proteins.

(c) Both ECM and cell walls are important in maintaining cell shape and in retaining water, thereby resisting compression.

(d) Roles unique to the ECM include regulation of cellular processes such as adhesion, motility, and differentiation during embryonic development. Roles unique to cell walls include protection of the cell from mechanical injury and microbial invasion, as well as provision of the mechanical support necessary to withstand the turgor pressure that gives plant tissues their rigidity.

11-2. (a) The ECM is held in place mainly by linkages of the proteoglycans to the proteins in the plasma membrane and by linkages between proteoglycans, collagen, and membrane receptors, mediated by adhesive glycoproteins. Antibodies against fibronectin and/or laminin, the two most common adhesive glycoproteins, would be useful in distinguishing between the two types of linkages. Examining the cross-linked peptides that remain after selective digestion of proteins might also provide information about the proteins involved in the crosslinks and about the chemical nature of the crosslinks as well.

(b) The adhesive glycoproteins bind proteoglycans and collagen molecules to each other and to receptors on the membrane surface, thereby reinforcing the linkage of collagen fibers to the surrounding matrix and the linkage of the ECM to the plasma membrane.

(c) To identify the binding properties of the various domains, biochemists cleaved the fibronectin and laminin molecules at specific sites along the polypeptide chain, isolated the individual segments, and assayed them in vitro for binding activity.

11-3. (a)   Collagen/elastin: Collagen is responsible for the strength of the ECM, whereas elastin imparts elasticity and flexibility to the ECM.

(b)   Fibronectin/laminin: These are the two most common kinds of adhesive glycoproteins; fibronectins occur widely throughout supporting tissues and body fluids, whereas laminins are found mainly in the basal laminae.

(c)   Integrin/selectin: The integrins are transmembrane proteins that serve as receptors for ECM proteins such as collagen, fibronectin, and laminin; selectin is the best-characterized integrin.

(d)   IgSF/cadherin: Both IgSFs and cadherins mediate cell-cell recognition and adhesion; the two groups of proteins can be distinguished from each other functionally because of the calcium requirement of cadherins but not IgSFs.

(e)   ECM/glycocalyx: The glycocalyx is a carbohydrate-rich zone juxtaposed between the plasma membrane and the ECM of many types of animal cells.

(f)   Hemidesmosome/focal adhesion: Both are adhesive junctions that connect cells to the ECM, with integrins as their main transmembrane linker proteins.

(g)   Apical surface/basolateral surface: The surfaces of intestinal epithelial cells that face the lumen of the intestine and the circulatory system, respectively; maintained as separate domains with respect to integral membrane proteins because tight junctions prevent proteins on the apical surface from moving laterally in the membrane to the basolateral surface, and vice versa.

11-4. Compaction is probably mediated by cadherins, since the proteins involved appear to be calcium-sensitive cell surface proteins. In reality, the molecule involved is E-cadherin, which was also "uvomorulin" in the early literature.

11-5. (a)   A   (c)   G   (e)   G, P

(b)   T   (d)   T

11-6. (a)   G; forms channels (cytoplasmic connections) between adjacent animal cells.

(b)   A; anchors actin microfilaments to the plasma membrane.

(c)   A; probably binds the plasma membranes of the two cells together.

(d)   P; tubular structure found in the center of a plasmodesma.

(e)   A; major proteins of the desmosome plaque.

(f)   P; forms channels (cytoplasmic connections) between adjacent plant cells.

(g)   A; an example of a transmembrane linker protein

(h)   T; binds adjoining cells together, creating fused ridges

11-7. (a) Primary cell walls are laid down as new cells are formed and are extensible under the appropriate conditions. In some plant cells, secondary cell walls are deposited on the inner surface of the primary wall. Because of their high cellulose and lignin contents, secondary cell walls are inextensible and therefore specify the final size and shape of the cell definitively.

(b) Cellulose comprises the rigid microfibrils of the cell wall, whereas hemicellulose is one of the major polysaccharides that make up the amorphous matrix in which the cellulose microfibrils are embedded.

(c) Xylan is a polymer of xylose, whereas xyloglucan is a polymer of glucose with side chains that consist of xylose, galactose, arabinose, and fucose. Xylan is found in monocots and xyloglucan in dicots.

(d) Extensins are rigid, rodlike glycoproteins with numerous oligosaccharide side chains that are tightly integrated into the polysaccharides matrix of the cell wall. Lignins are polymers of aromatic alcohols that form cross-linked networks with extensin and other cell wall components.

(e) The desmotubule is the cylindrical structure in the central channel of a plasmodesma, and the annulus is the ring of cytoplasm between the desmotubule and the membrane that lines the desmotubule.

(f) The plasmodesma is the junction that connects the cytoplasms of two adjacent plant cells; it plays the same role as the gap junction between animal cells.

11-8. (a) Hydroxyproline residues, but apparently not unhydroxylated proline residues, stabilize the collagen triple helix. (The hydroxyl groups of this amino acid form interchain hydrogen bonds that help stabilize the assembled triple-stranded helix.) If Vitaminc C (ascorbic acid) is deficient in the diet, tissue levels of ascorbic acid will be low and the enzyme prolyl hydroxylase is not maintained in the reduced form and is therefore inactive. Proline is not hydroxylated, resulting in the inadequate stabilization of the triple helix. Collagen therefore breaks down, leading to defects in tissues that depend on collagen to maintain the adhesive strength of the ECM. As a consequence, such tissues are subject to breakdown and bruising. (Blood vessels also become fragile, which accounts for the hemorrhaging that is characteristic of scurvy.)

(b) For the disease condition to become progressively worse—and for a patient with scurvy to respond to dietary vitamin C—collagen degradation and replacement must be relatively rapid, at least in susceptible tissues. (The turnover rate of collagen varies greatly, in fact; it is quite rapid in connective tissue but very slow in bone.)

(c) Sailors on long sea voyages were susceptible to scurvy because no fresh fruit—and hence no source of vitamin C—was available. This is no longer a problem because once the connection between vitamin C deficiency and scurvy was established, sailors were provided with citrus fruit. The term "limey" reflects the British Navy's use of limes for this purpose.

11-9.   (a)   *Hypothesis:* Adjacent cells in tissues are in direct electrical connection with each other, presumably by means of channels that interconnect adjacent cells.

Conclusion: Direct electrical connections exist between cells and, since current in living cells is carried by small ions, cells must be connected by channels (now called gap junctions) that permit ions to flow directly from the cytoplasm of one cell to the cytoplasm of an adjacent cell without leaving either cell.

Supportive findings: Current flows much more readily between adjacent cells than between either cell and outside of the cell.

Further experimental questions: Is the existence of direct electrical connection between adjacent cells common to most cells in most tissues or just to specialized cell types? Can such electrical connections be demonstrated between numerous cells in a row? What is the size of such channels?

(b)   *Hypothesis:* Gap junctions between cells will allow passage of molecules and ions up to some specific size limit.

Conclusion: The gap junctions that connect adjacent cells in insect salivary glands allow the passage of small molecules (molecular weight up to at least 1158 daltons), whereas larger molecules (1926 daltons) do not flow between cells.

Supportive findings: When injected into a cell, fluorescent molecules with a molecular weight of 1158 daltons appear in adjacent cells but molecules with a molecular weight of 1926 daltons do not, as Figure 11-29 in the textbook illustrates.

Further experimental questions: Is there a finite cut-off between 1158 and 1926 daltons in the case of insect salivary gland cells? Is the size limit the same for gap junctions between cells in other tissue types? Is the cut-off point dependent on the chemical nature of the fluorescent molecule? Are ions subject to the same size restrictions and cut-off points?

(c)   *Hypothesis:* The passage of molecules between cells can be regulated by the intracellular Ca^{2+} content.

Conclusion: When the Ca^{2+} concentration is increased in an individual cell, the movement of fluorescent molecules into that cell is inhibited, suggesting that gap junctions are closed under these conditions.

Supportive findings: Providing that the plasma membranes of the cells are intact, fluorescent molecules migrate from one cell to another even though the experiment is carried out in a Ca^{2+} -containing medium. If the plasma membranes of cells are punctured to allow calcium from the medium to enter the cells, fluorescent molecules do not move into those cells, as Figure 11-26b in the textbook shows.

Further experimental questions: Is this effect specific to calcium ions? Is the closing of gap junctions an all-or-nothing or can varying degrees of permeability be demonstrated at intermediate Ca^{2+} concentrations? Can other chemical factors alter gap junction permeability?

12

Intracellular Compartments: The Endoplasmic Reticulum, Golgi Complex, Endosomes, Lysosomes, and Peroxisomes

12-1. (a) Both peroxisomes and mitochondria. The process of β oxidation of long chain fatty acids begins in the peroxisome but once the fatty acids have been shortened sufficiently in chain length, their further oxidation occurs in the mitochondrion. In the peroxisome, β oxidation generates NADH and acetyl CoA, which are essential for other peroxisomal reactions. In the mitochondrion, the acetyl CoA is immediately available for further oxidation by the TCA cycle.

(b) ER. Cholesterol is synthesized in the ER, where it is immediately available for steroid hormone biosynthesis or is easily transported throughout the endomembrane system.

(c) Rough ER and Golgi. Once in the ER, insulin can enter the secretory pathway for export from the cell.

(d) Smooth ER. Once in the ER, testosterone can enter the secretory pathway for export from the cell.

(e) Lysosome. Because the lytic enzymes that are needed for autophagy are sequestered in a membrane-bounded organelle, other cellular components are protected from degradation.

(f) Rough ER and Golgi. Membrane proteins can be glycosylated by the same general pathway used for glycosylating proteins destined for different locations within the cell, with variations in glycosylation used as targeting signals.

(g) Smooth ER. Phenobarbital is hydrophobic and therefore accumulates in the lipid portion of the smooth ER membrane; hydroxylation increases the solubility of the molecule in water, allowing the cell to export it via a secretory pathway.

(h) Golgi. The proteins are sorted after undergoing glycosylation, folding, and additional processing steps, thereby enabling the cell to use the same general pathway for processing proteins that are then targeted to different organelles.

12-2. (a   R, S   (d)   S   (g)   S

      (b)   R   (e)   R   (h)   R

      (c)   S   (f)   R   (i)   R, S

12-3. (a)   For the pathway whereby glycoproteins of the plasma membrane are synthesized and glycosylated, see Figure S12-1.

      (b)   Glycoproteins are always found on the outer phospholipid monolayer of the plasma membrane with their carbohydrate side chains exposed on the outer surface because this is the monolayer that originally faced the interior of the rough ER and Golgi, where the enzymes involved in glycosylation are located.

      (c)   The answer to parts a and b require the assumptions that (i) membrane asymmetry is maintained throughout the rough ER, Golgi, and plasma membrane, and (ii) the relationship of these membrane systems to each other is essentially as illustrated in Figure S12-1.

12-4. (a)   C   (d)   C, I, II   (g)   I   (j)   C, I, II

      (b)   II   (e)   C   (h)   C   (k)   I, II

      (c)   I   (f)   N   (i)   C

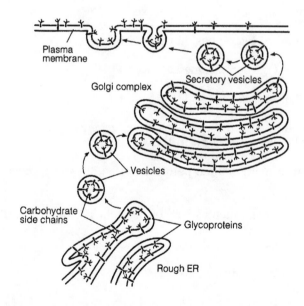

Figure S12-1   Synthesis and Glycosylation of Integral Membrane Proteins of the Plasma Membrane. Integral membrane proteins are synthesized on the rough ER, with oligosaccharide side chains added in part on the lumenal side of the rough ER (core glycosylation) and in part on the lumenal side of the Golgi complex (terminal glycosylation). Side chains therefore face the interior of both organelles as well as the interior of transport vesicles, and become oriented toward the exterior of the cell when the vesicles fuse with the plasma membrane. See Problem 12-3.

12-5. (a) Colcichine is a drug that prevents microtubule assembly, thereby disrupting microtubule-based functions (see Chapter 22). This observation indicates the involvement of microtubules in intracellular movement of exocytic and endocytic vesicles.

(b) This finding suggests that the pathways for constitutive and regulated secretion, though separate, can occur simultaneously in the same cell.

(c) This result suggests that exocytic secretion may be triggered in vivo by an elevation of the intracellular Ca^{2+} concentration.

(d) This observation indicates that dynamin is essential for receptor-mediated endocytosis. However, there are also dynamin-independent pathways for ingesting extracellular fluid. When receptor-mediated endocytosis is inhibited, the rate of ingestion of extracellular fluid by other endocytic pathways increases.

12-6.

(a)	P, R, A, E	(d)	A	(g)	P, R, A, E	(j)	P, R, A
(b)	A	(e)	E	(h)	P, R	(k)	P, R, A
(c)	P, R, E	(f)	P, A, E	(i)	E		

12-7. (a) N; peroxisomes acquire proteins from ribosomes found in the cytosol.

(b) S; true only of glyoxysomes and some animal peroxisomes.

(c) N; acid hydrolases are found in Iysosomes, not peroxisomes.

(d) A; catalase is a component of all peroxisomes.

(e) A; general property of peroxisomes.

(f) N; fireflies and related bioluminescent organisms have luciferase protein present in their peroxisomes but DNA is not found in this organelle.

(g) S; urate oxidase is present in many animal peroxisomes and in some, but not all, plant peroxisomes.

(h) A; peroxisomes are a source of as much as 50% of the dolichol found in cells.

(i) A; general property of peroxisomes.

12-8.

(a)	M	(c)	S	(e)	S	(g)	M
(b)	S	(d)	N	(f)	S	(h)	N

12-9. (1) For the short peptide Lys-Asp-Glu-Leu:

(a) Incorporated into ER-specific proteins as the polypeptides are synthesized by ribosomes attached to the rough ER.

(b) When such proteins reach the Golgi complex, they bind to specific receptors and are packaged into transport vesicles for return to the ER.

(c) Removal of this tag from ER-specific proteins usually results in their secretion from the cell.

(2) For hydrophobic membrane-spanning domains:

(a) Incorporated into membrane-bound Golgi complex proteins as the polypeptides are synthesized by ribosomes attached to the rough ER.

(b) Such proteins tend to move through the endomembrane system until the thickness of the membrane, which increases progressively from the ER to the plasma membrane, exceeds the length of their membrane-spanning domains and blocks further migration.

(c) Removal of the membrane-spanning domains would convert the membrane-bound proteins to soluble proteins and would also severely affect folding of the polypeptides; such abnormal proteins will probably be exported from the ER and degraded.

(3) For mannose-6-phosphate residues:

(a) Mannose-6-phosphate (M-6-P) residues are attached to soluble lysosomal proteins by the sequential action of several enzymes found in the lumen of the rough ER and Golgi complex.

(b) When such proteins reach the TGN, they bind to M–6–P receptors and are packaged into vesicles for transport to endosomes.

(c) Removal of M-6-P residues from lysosomal proteins often leads to secretion of the proteins; however, evidence suggests that additional tags may ensure delivery of some lysosomal proteins to endosomes.

12-10. (a) The fibers or particles are probably taken up by endocytosis, followed by transport via early and late endosomes to a heterophagic lysosome.

(b) The fibers or particles may physically abrade the lysosomal membrane, causing it to become leaky.

(c) Cell death is probably due to the digestion of cellular components by acid hydrolases that escape from damaged lysosomes.

(d) The fibers or particles released on cell death presumably are available for ingestion by other macrophages, with the same end result. Because the fibers and particles are not digestible and there is no mechanism to remove them from the lungs, a cycle of uptake, lysosomal damage, cell death, and fiber or particle release is set up that can continue indefinitely, killing more and more cells.

(e) Exposure of silica particles apparently causes the macrophages to release a soluble factor that stimulates fibroblast cells in the lung to deposit collagen fibers, probably in an attempt to seal off and thereby contain the silica in the lung.

12-11. (a)   Proteins and RNA molecules have the lowest S values, whereas nuclei have the largest S values. The particles with the lowest and highest densities are chloroplasts and RNA molecules, respectively.

(b)   The density of a ribosome (range: 1.5-1.6 g/cm^3) is intermediate between those for RNA (about 2.0 g/cm^3) and proteins (about 1.3 g/cm^3). This is reasonable because a ribosome contains both RNA and protein molecules and therefore has a density that represents the weighted average of its components.

(c)   Whereas the *density* of a virus represents the weighted average of its components, its *S value* is related, though not linearly, to its mass, which is obviously the sum of the masses of its components.

(d)   Yes; differential centrifugation separates particles based on differences in S values, and those for nuclei (between 10^6 and 10^7) and for mitochondria (between 10^4 and 10^5) are sufflciently different (by at least one, more likely two or three, orders of magnitude) to allow effective separation of these two kinds of organelles. However, the range of S values for peroxisomes and mitochondria overlap almost entirely, so differential centrifugation cannot be used to separate these two kinds of organelles.

(e)   Equilibrium density centrifugation cannot be used to separate microsomes from mitochondria because the technique depends entirely on density differences and these two kinds of organelles are not significantly different from each other in density. In fact, the density range for mitochondria falls entirely within the range for microsomes. However, the technique is very effective for the resolution of RNA (density of about 2.0 g/cm^3) from DNA (density of about 1.7 g/cm^3).

(f)   The separation scheme depends on differential centrifugation to isolate lysosomes, mitochondria, and peroxisomes from other cellular constituents, followed by equilibrium density centrifugation to separate Iysosomes, mitochondria, and peroxisomes from each ,other. For details of these procedures, see Figures 12-4 and 12A-7 in the textbook. For the use of the detergent WR–1339 as a means of decreasing the density of Iysosomes enough to allow resolution of lysosomes and mitochondria on an equilibrium density gradient despite their initial overlap in densities, see Figure, 12-23 in the textbook. Alternatively, density gradient centrifugation (Figure 12A-5 in the textbook) could be used to resolve these two organelles from each other following equilibrium density centrifugation.

12-12. (a)   Minimal criteria would be (1) a set of enzymes that is different from that of any other type of organelles and could therefore be used as marker enzymes for the new organell, and (2) a means of separating the new set of enzymes (and hence the new organelle) from all other organelles or cell structures. The most generally useful means of achieving such separation is differential centrifugation followed by equilibrium density centrifugation.

(b)   Yes, provided that equilibrium density centrifugation is part of the procedure used to achieve separation.

(c)   Probably not, because peroxisomes and Iysosomes are usually very similar in size, shape, and density.

CHAPTER

13

Chemotrophic Energy Metabolism: Glycolysis and Fermentation

13-1.   (a)   F; energy is always *required* to break a covalent bond, including the phospho-anhydride bond that links the terminal phosphate group to the rest of the ATP molecule. ATP is a "high-energy" compound because its hydrolysis is exergonic, which means that more energy is released as the bonds between the –H and –OH groups of water are formed than is required to break the phosphoanhy dride bond of ATP.

   (b)   T

   (c)   T

   (d)   F; the phosphate group doesn't "possess" any intrinsic energy of its own. The term "high-energy" applies only to the phosphoanhydride bond that links the phosphate group to the rest of the ATP molecule, and then only in the sense explained in the answer to part a.

   (e)   F; "low-energy bonds" require *more* energy to break, which is why less energy is released when such bonds are hydrolyzed.

   (f)   T

13-2.   (a)   The clue is in the word *zymase,* an early term for enzyme. The heat-labile fraction (zymase) contains the enzymes and the heat-stable fraction (cozymase) contains the coenzyme (NAD^+) necessary for fermentative activity. This observation is important in distinguishing enzymes from coenzymes and in understanding the need for both.

   (b)   Phosphate is required as substrate in reaction Gly-6, and the sequence cannot function without it. This observation is important in establishing the involvement of phosphate groups in energy metabolism.

   (c)   Based on the accumulation of a doubly phosphorylated sugar, we can deduce that iodoacetate must be an inhibitor of aldolase, the enzyme that splits fructose-1,6-bisphosphate into two trioses. In the presence of iodoacetate, aldolase is inactive and fructose-1,6-bisphosphate accumulates. This observation is important to establish fructose-1,6-bisphosphate as an intermediate in the pathway. (Iodoacetate is, in fact, a general inhibitor of Mg^{2+}-requiring enzymes, of which aldolase is an example.)

(d)   Fluoride ion is a potent inhibitor of enolase, the enzyme that catalyzes the removal of water from 2-phosphoglycerate to generate phosphoenolpyruvate in reaction Gly-9. Addition of fluoride to the yeast extract caused the accumulation of two phosphorylated three-carbon acids, which could then be identified as 2-and 3-phosphoglycerate, thereby establishing the chemical nature of two further intermediates in the pathway.

13-3.   (a)   From phosphoenolpyruvate: exergonic

From glucose-1-phosphate: endergonic

From phosphocreatine: exergonic

From glycerol phosphate: endergonic

(b)   The reaction is as follows:

1,3-bisphosphoglycerate + ADP → ATP + 3-phosphoglycerate.

The $\Delta G^{\circ\prime}$ for the reaction is calculated as the difference in energy of hydrolysis between 1,3-bisphosphoglycerate and ATP:

$\Delta G^{\circ\prime} = -11.8 - (-7.3) = $ **–4.5 kcal/mol.**

(c)   The reactions are as follows:

Phosphoenolpyruvate + ADP → pyruvate + ATP $\Delta G^{\circ\prime} = -7.5 \text{ kcal/mol}$

Glucose + ATP → glucose-6-phosphate + ADP $\Delta G^{\circ\prime} = -4.0 \text{ kcal/mol}$

Because the hydrolysis of phosphoenolpyruvate (PEP) is a more highly exergonic reaction than is the hydrolysis of ATP, the transfer of a phosphate group from PEP to ADP will be thermodynamically spontaneous, and the presence of the enzyme pyruvate kinase ensures that it is kinetically feasible as well. Similarly, the transfer of a phosphate group from ATP to glucose is also thermodynamically spontaneous and is catalyzed by the enzyme hexokinase. (Note that the net reaction is the phosphorylation of glucose at the expense of PEP but that the phosphate group is transferred from PEP to glucose via ATP/ADP.)

(d)   The equilibrium mixture will contain ADP, ATP, PEP, pyruvate, glucose, and glucose-6-phosphate because each of these molecules is involved in the reactions of part c and all will therefore be present at equilibrium.

(e)   Yes, 1,3-bisphosphoglycerate (BPG) is also a possible phosphate donor for the generation of glycerol phosphate because BPG has a higher free energy of hydrolysis under standard conditions than does glycerol phosphate. However, this reaction does not occur in cells (that is, there is no known enzyme that catalyzes the direct transfer of a phosphate group from BPG to glycerol), probably because the reaction would be too highly exergonic and would release too much energy as heat.

(f)   *Yes* for the formation of glycerol phosphate but *no* for the formation of glucose-1-phosphate. The transfer of a phosphate group from BPG to ADP is sufficiently exergonic ($\Delta G' = -4.5$ kcal/mol) to drive the phosphorylation of glycerol ($\Delta G' = +2.2$ kcal/mol) but does not yield enough energy to drive the phosphorylation of glucose on the #1 carbon atom ($\Delta G' = +5.0$ kcal/mol).

13-4.   (a)   glycolysis is the first stage in aerobic energy metabolism for any cell that depends on glucose as its energy source, as brain cells do.

(b)   the oxidation that occurs at one step is balanced by the reduction that occurs when pyruvate is converted to lactate or ethanol and carbon dioxide.

(c)   whether an external electron acceptor is available, which in most cases means whether oxygen is available (i.e., whether the cell is functioning under aerobic conditions).

(d)   the pyruvate that is then reductively decarboxylated to generate the ethanol present in the beer and the carbon dioxide that causes the bread to rise.

(e)   the liver and heart muscle.

(f)   energy is always lost (as heat and entropy) whenever a reaction occurs, so glucose formation from lactate requires more energy than glucose catabolism releases.

13-5.   (a)   The glycosidic bond that links successive glucose units together in a polysaccharide has sufficient free energy of hydrolysis to allow it to be cleaved by phosphorolysis with the direct uptake of inorganic phosphate. The product is therefore a glucose molecule that is already phosphorylated (on carbon atom 1), which means that step Gly-1 of the glycolytic pathway is bypassed and the ATP that would otherwise be required there is saved.

(b)   In the intestine, sucrose is hydrolyzed to free glucose and fructose, each of which is then catabolized by the glycolytic pathway with the expected yield of 2 molecules of ATP per molecule of monosaccharide.

(c)   Because monosaccharide units are cleaved from a polysaccharide by phosphorolysis, we can suggest the same mechanism for bacterial sucrose catabolism. Phosphorolysis of a disaccharide will generate one phosphorylated monosaccharide that will yield three molecules of ATP by the glycolytic pathway (because step Gly-1 is bypassed and one less ATP molecule is needed) and one free monosaccharide that will require step Gly-1 and will therefore yield two molecules of ATP. In fact, the sucrose phosphorylase reaction generates glucose-1-phosphate and free fructose, and the glycolytic ATP yield is therefore 5 molecules of ATP per molecule of sucrose, or 2.5 ATP molecules per monosaccharide.

(d)   Raffinose has three monosaccharides linked together by two glycosidic bonds. Phosphorolysis of these bonds generates two phosphorylated hexoses (yield: 3 ATP each) and one free hexose (yield: 2 ATP). The average ATP yield per monosaccharide is therefore $(3 + 3 + 2)/3 = \mathbf{2.67}$.

13-6. (a) $\Delta G' = \Delta G^{\circ\prime} + RT \ln \dfrac{[\text{glucose - 6 - phosphate}]}{[\text{glucose}]\,[\text{P}_i]}$

$= +3300 + (1.987)\,(298) \ln \dfrac{0.08 \times 10^{-3}}{[\text{glucose}](1.0 \times 10^{-3})}$

$= +3300 + 592 \ln \dfrac{0.08}{[\text{glucose}]}$

$= +3300 + 592 \ln (0.08) - 592 \ln [\text{glucose}]$

$= +3300 - 1495 - 592 \ln [\text{glucose}]$

At equilibrium, $\Delta G' = 0$, so $\Delta G' = 1805 - 592 \ln [\text{glucose}] = 0$.

Because $\ln [\text{glucose}] = \dfrac{1805}{592} = 3.049$, $[\text{glucose}] = e^{3.049} = \boldsymbol{21\ M!}$

This means that a glucose concentration of 21 M would be required just to bring the reaction to equilibrium; anything over this would render the reaction spontaneous in the direction of phosphorylation. This is impossible; even 2 M glucose would be a thick syrup!

(b) Glucose + ATP $\longrightarrow$ glucose-6-phosphate + ADP

$\Delta G^{\circ\prime} = +3.3\ \text{kcal/mol} - 7.3\ \text{kcal/mol} = \boldsymbol{-4.0\ kcal/mol}$.

(c) $\Delta G' = \Delta G^{\circ\prime} + RT \ln \dfrac{[\text{glucose - 6 - phosphate}]\,[\text{ADP}]}{[\text{glucose}]\,[\text{ATP}]}$

$= -4000 + (1.987)\,(298) \ln \dfrac{(0.08 \times 10^{-3})(0.15 \times 10^{-3})}{[\text{glucose}](1.8 \times 10^{-3})}$

$= -4000 + 592 \ln \dfrac{(6.67 \times 10^{-6})}{[\text{glucose}]}$

$= -4000 + 592 \ln (6.67 \times 10^{-6}) - 592 \ln [\text{glucose}]$

$= -4000 - 7055 - 592 \ln [\text{glucose}] = 11{,}055 - 592 \ln [\text{glucose}]$

At equilibrium, $\Delta G' = 0$, so $\Delta G' = -11{,}055 - 592 \ln [\text{glucose}] = 0$. Because $\ln [\text{glucose}] = -11{,}055/592 = -18.67$, $[\text{glucose}] = e^{-18.67} = \boldsymbol{7.76 \times 10^{-9}\ M!}$ This means that a glucose concentration of 7.76×10^{-9} M would bring the reaction to equilibrium; any glucose concentration higher than this will render the reaction spontaneous in the direction of phosphorylation. This is physiologically very reasonable, because glucose phosphorylation is thermodynamically feasible as long as the glucose concentration remains more than 0.01 μM.

(d) From 2.1×10^{1} to 7.76×10^{-9} is more than nine orders of magnitude!

(e) $\Delta G' = \Delta G^{\circ\prime} + RT \ln \dfrac{\left[\text{glucose - 6 - phosphate}\right]\left[\text{ADP}\right]}{\left[\text{glucose}\right]\left[\text{ATP}\right]}$

$= -4000 + 592 \ln \dfrac{(0.08 \times 10^{-3})(0.15 \times 10^{-3})}{(5.0 \times 10^{-3})(1.8 \times 10^{-3})}$

$= -4000 + 592(-6.623)$

$= -4000 - 3920 = -7920^{\text{cal/mol}} = \textbf{-7.9 kcal/mol}.$

13-7.   (a)   Ethanol catabolism in the body begins with its oxidation (dehydrogenation), with NAD^+ as the electron acceptor. The more ethanol that is consumed, the greater the demand for NAD^+ and the more serious the reduction in NAD^+ concentration, which means, in turn, that the supply of NAD^+ may be inadequate for aerobic respiration of glucose.

(b)   Acetaldehyde is the immediate product of ethanol oxidation:

Ethanol + NAD^+ → acetaldehyde + NADH + H^+

(c)   Methanol and ethanol are both substrates of the enzyme alcohol dehydrogenase and therefore compete for the active site. The body is flooded with a large amount of ethanol to provide an effective competitor of methanol, thereby minimizing the production of formaldehyde and lessening the danger that the patient will be "pickled."

13-8.   (a)   Propionate differs from pyruvate in that its middle carbon atom is at the hydrocarbon level rather than at the carbonyl level. To reduce pyruvate to propionate would therefore require two molecules of NADH, assuming an appropriate sequence of reactions exists. But the stoichiometry of glycolysis provides only one molecule of NADH per molecule of pyruvate, not two. Therefore, all of the pyruvate cannot be reduced to propionate.

(b)   If 50% of the pyruvate molecules are reduced to propionate and the remaining 50% are left as pyruvate, the stoichiometry will come out right. The overall reaction would then be

$C_6H_{12}O_6 \longrightarrow CH_3\text{–}CO\text{–}COO^- + CH_3\text{–}CH_2\text{–}COO^- + H_2O + 2\,H^+$

Glucose   Pyruvate   Propionate

(c)   The pyruvate must be further metabolized by decarboxylation.

13-9.   (a)   Galactose + ATP $\longrightarrow$ glucose-6-phosphate + ADP.

(b)   Quite similar, because the net result in both cases is an ATP-linked phosphorylation of a six-carbon sugar, and thermodynamic parameters are independent of route.

(c)   The reaction sequence for the epimerase reaction is shown in Figure S13-1.

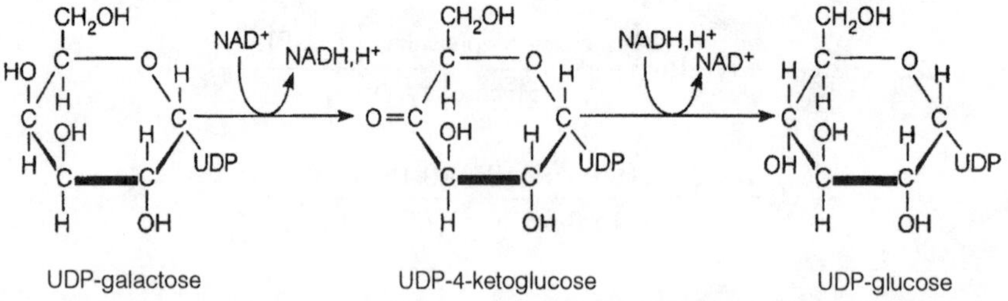

Figure S13-1   The Epimerization of UDP-Galactose to UDP-Glucose. The conversion of UDP-galactose to UDP-glucose is a two-step epimerization sequence in which the UDP-bound galactose is first oxidized by NAD^+ to the corresponding keto compound, UDP-4-ketoglucose, which is then reduced by NADH to UDP-glucose. See Problem 13-9.

(d)   The inability to catabolize galactose leads to galactose accumulation in the tissues and blood of an organism that continues to ingest, absorb, and hydrolyze lactose to get the glucose it needs for energy metabolism. The elevated level of galactose in the blood is apparently deleterious to brain and lens cells.

(e)   Galactosemia can also be caused by the genetic absence of the epimerase enzyme that catalyzes reaction 13-25 in the textbook. The absence of this enzyme causes a buildup of UDP-galactose, presumably leading to an elevated level of galactose in the blood, resulting in the deleterious effects described in part d.

13-10. (a)   A reaction sequence that is thermodynamically feasible (exergonic) in one direction will not function in the other direction by simple reversal of each of the reactions because it will be endergonic in that direction under the same conditions. (Theoretically, one can imagine conditions in which a sufficiently high concentration of product(s) and a sufficiently low concentration of reactant(s) might drive a reaction sequence in the opposite direction, but the thermodynamic driving force—that is, the $\Delta G'$ for the reaction sequence in the forward direction—is usually sufficiently great for most metabolic pathways that it is virtually impossible to reverse such pathways simply by changes in reactant and/or product concentrations.)

(b)   2 pyruvate + 4 ATP + 2 GTP + 6 H_2O + 2 NADH + 2 $H^+ \rightarrow$

glucose + 4 ADP + 2 GDP + 6 P_i + 2 NAD^+

(c)   The four additional phosphoanhydride bonds provide the extra energy needed to ensure that the pathway is driven strongly enough in the gluconeogenic direction so that it is essentially irreversible in that direction.

(d)   The $\Delta G'$ for the glycolytic pathway under typical cellular conditions is about –20 kcal/mol and therefore + 20 kcal/mol in the opposite direction. Because four additional phosphoanhydride bonds are hydrolyzed to drive the pathway in the opposite direction and each of those bonds has a $\Delta G'$ of about –10 kcal/mol, the net driving force in the gluconeogenic direction is:

$$\Delta G' = 20 + 4(-10) = \textbf{–20 kcal/mol.}$$

(e)   The key glycolytic and gluconeogenic enzymes shown in Figure 13-13 in the textbook are subject to regulation by a variety of factors, including the AMP, ADP, ATP, acetyl CoA, and F2,6BP status of the cell. Changes in the concentrations of these intermediates either activate or inhibit the regulatory enzymes, thereby effectively turning the pathway "on" in one direction and "off" in the other.

13-11. (a)   The discovery that glycolytic enzymes are compartmentalized in trypanosomes came about as a result of differential centrifugation, a technique that causes most organelles to sediment to the bottom of a centrifuge tube in response to centrifugal force while molecules, ions, and smaller organelles such as ribosomes remain in the supernatant. In an experiment in which glycolytic enzymes were expected to remain in the supernatant, seven out of the ten enzymes did not, but were recovered in the pellet instead, indicating an organellar location.

(b)   With the glycolytic enzymes compartmentalized in this way, the enzymes and intermediates in the pathway can be maintained in relatively high concentrations because the organellar volume is small, compared to the volume of the cell.

(c)   The first seven steps of the glycolytic pathway begin with glucose and lead to the formation of 3-phosphoglycerate. The glycosomal membrane must therefore have transport proteins for glucose (which moves inward), 3-phosphoglycerate (which moves outward), and inorganic phosphate (which must move inward to balance the loss of phosphate loss as 3-phosphoglycerate moves outward). (Bonus question: Can you suggest a means whereby the outward movement of 3-phosphoglycerate can be coupled to the inward movement of inorganic phosphate?)

(d)   By biochemical definition, a peroxisome is capable of carrying out the generation and degradation of hydrogen peroxide (H_2O_2) and always possesses at least one H_2O_2-generating enzyme (an oxidase) and one H_2O_2-degrading enzyme (catalase).

13-12. (a)   The ATP yield per unit of glucose catabolized is much lower under anaerobic conditions (2 ATP/glucose) than under aerobic conditions (36-38 ATP/glucose). A yeast cell functioning under anaerobic conditions must therefore consume 18 to 19 times as much glucose per unit time as under aerobic conditions in order to sustain the same rate of ATP generation.

(b)   Although the glycolytic pathway responsible for glucose fermentation by yeast cells begins and ends with an unphosphorylated molecule (glucose and ethanol, respectively), almost all of the intermediates in the process are phosphorylated compounds. The fermentation process therefore requires inorganic phosphate, which is taken up in step Gly-6 and used to generate ATP from ADP in step Gly-7, and will be stimulated by the addition of inorganic phosphate if phosphate is a limiting reagent in the culture medium.

(c)   The ATP needed for the rapid but short-term movements of an alligator is generated by glycolysis at the expense of muscle glycogen. The long period of recovery following such movements is needed for the replenishment of muscle glycogen stores.

(d)   The oxidation of glyceraldehyde-3-phosphate that occurs at step Gly-6 involves the oxidation of a carbonyl group to a carboxylic acid group, which is a highly exergonic reaction. The reduction of pyruvate to lactate is endergonic but involves the reduction of a carbonyl group to a hydroxyl group, a reaction that requires less energy than is released by the oxidation of a carbonyl group to a carboxylic acid group. (Note, then, that anaerobic life is possible only because of the difference in energy levels between a carboxyl group and a hydroxyl group!)

13-13. (a)   It allows substrate oxidation to proceed without concomitant ATP generation, releasing this step in the glycolytic pathway from its normal sensitivity to, and regulation by, the availability of ADP and P_i.

(b)   Use of arsenate instead of phosphate at step Gly-6 results in spontaneous hydrolysis of the arseno intermediate without conservation of the energy of the bond as ATP. This results in two molecules less of ATP per molecule of glucose, so the energy yield under anaerobic conditions is zero, and the arsenate is therefore fatal.

c)   Any reaction involving the direct uptake of inorganic phosphate that leads to the generation of a high-energy phosphate bond and the formation of ATP may be subject to uncoupling in this way, provided only that the enzyme will accept arsenate in place of phosphate at its active site.

13-14. (a)   Fructose-2,6-bisphosphate (F2,6BP) is an allosteric activator of phosphofructo-kinase (PFK). The enzyme shows normal Michaelis-Menten kinetics in the presence of F2,6BP (red line) but not in its absence (black line). In fact, the apparent K_m of the enzyme for fructose-6-phosphate is at least five times higher in the absence of F2,6BP than in its absence. F2,6BP, in other words, is required for normal functioning of PFK. (The sigmoidal dependence of reaction rate of substrate concentration in the absence of F2,6BP is characteristic of an allosterically regulated enzyme when functioning in the absence of its allosteric effector.)

(b)   ATP is an allosteric inhibitor of PFK. The enzyme shows normal Michaelis-Menten kinetics when the ATP concentration is low (black line) but not when the ATP concentration is high (red line). In fact, the apparent K_m of the enzyme for fructose-6-phosphate is at least five times higher in the presence of a high concentration ATP than it is in when the ATP concentration is low. In other words, the ATP concentration must be kept low for PFK to function normally. (Note, however, that ATP is not only an allosteric effector of PFK but also one of its substrates, so normal functioning of the enzyme requires an ATP concentration that is low enough to avoid allosteric inhibition but high enough to sustain catalytic activity at the active site.)

(c)   In Figure 13-16a, we must assume that the ATP concentration is high enough to sustain catalytic activity (i.e., significantly greater than the K_m of the enzyme for ATP) but still low enough to avoid allosteric inhibition by ATP. In Figure 13–16b, we must assume that the F2,6BP concentration is high enough (0.13 mM, for example) to ensure allosteric activation of the enzyme.

14

Chemotrophic Energy Metabolism: Aerobic Respiration

14-1. (a) M   (d) IM   (g) IM   (j) IS
       (b) IM   (e) M   (h) NO
       (c) IS   (f) OM   (i) IM

14-2. (a) C   (d) PM   (g) NO   (j) EX
       (b) PM   (e) C   (h) C
       (c) C   (f) PM   (i) PM

14-3. (a) F; the orderly flow of carbon through the TCA cycle is possible because most (all but one) of the enzymes of the cycle are present in soluble form in the mitochondrial matrix, where the substrate(s) of each enzyme can collide with, and bind specifically to, the active site of the enzyme.

       (b) T

       (c) F; respiration is an aerobic process in many organisms because oxygen is the single most common (though by no means the only) electron acceptor for reoxidation of reduced coenzymes.

       (d) T

       (e) T

       (f) F; only eight cycles of β oxidation are required.

14-4. (a) In; 2   (e) No flux   (j) No flux
       (b) In; 6   (f) No flux   (j) Out; 42
       (c) Out; 36   (g) No flux   (k) In; 2 pairs per glucose
       (d) In; 36   (h) No flux   (l) In and out; no net flux

14-5. (a) The pathway from citrate to α-ketoglutarate is as shown in Figure S14-1.

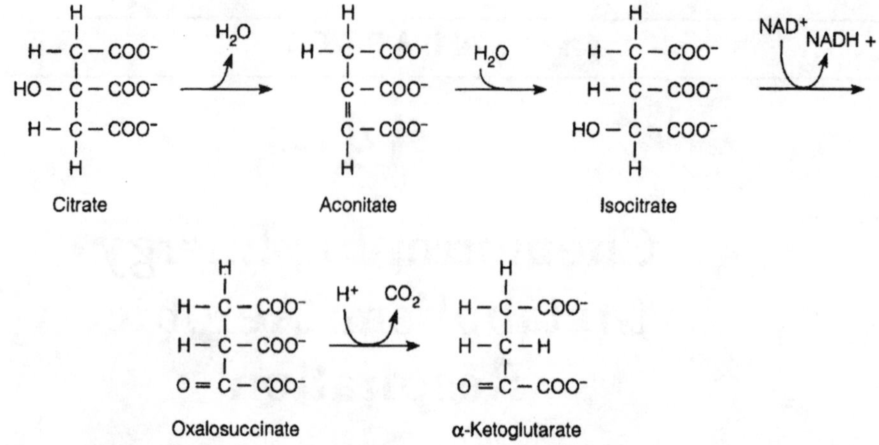

Figure S14-1 The Metabolic Path from Citrate to α-Ketoglutarate. The conversion of citrate into α-ketoglutarate proceeds in a four-step sequence involving dehydration, hydration, oxidation and decarboxylation reactions. See Problem 14-5.

(b)   The pathway from pyruvate and alanine to glutamate is as follows:

Pyruvate + CO_2 + ATP + H_2O ⟶ oxaloacetate + ADP + P_i

Pyruvate + CoA–SH + NAD^+ ⟶ acetyl CoA + CO_2 + NADH + H^+

Acetyl CoA + oxaloacetate + H_2O ⟶ citrate + CoA–SH

Citrate ⟶ isocitrate

Isocitrate + NAD^+ ⟶ α-ketoglutarate + CO_2 + NADH + H^+

α-Ketoglutarate + alanine ⟶ glutamate + pyruvate

Pyruvate + alanine + ATP + 2 H_2O + 2 NAD^+ ⟶

                    glutamate + CO_2 + ADP + P_i + 2 NADH + 2 H^+

(c)   By following the middle carbon atom of pyruvate in Figure 14-8 in the textbook, you should be able to convince yourself that it becomes the carboxyl carbon of acetate and hence the newly added ("upper") carboxyl group in citrate.

14-6. (a)   Yes, isocitrate can pass electrons exergonically to NAD^+ under standard conditions because the α-ketoglutarate/isocitrate redox pair has a more negative E_0' (–0.38V) than does the NAD^+/NADH redox pair (–0.32V), which means that the reduced form of the isocitrate/α-ketoglutarate redox pair (i.e., isocitrate) will spontaneously reduce the oxidized form of the NAD^+/NADH redox pair (i.e., NAD^+).

(b)   The E_0' value for the reduction of NAD^+ by isocitrate is calculated as:

$$\Delta E_0' = E_0', \text{ acceptor} - E_0', \text{ donor} = -0.32V - (-0.38V) = -0.32 + 0.38 = \textbf{+0.06}V.$$

The $\Delta E_0'$ value is positive, so the transfer of electrons from isocitrate to NAD^+ is thermodynamically spontaneous under standard conditions, as predicted in part a.

(c)   The $\Delta G^{o'}$ value for the reduction of NAD^+ by isocitrate is calculated as:

$$\Delta G^{o'} = -nF\Delta E_0' = -1(23{,}062)(+0.06) = -1394^{\text{cal/mol}} = \textbf{-1.38 kcal/mol.}$$

This is relevant to aerobic energy metabolism because the oxidation of isocitrate to α-ketoglutarate is one of the reactions of the TCA cycle (Reaction TCA-3).

(d)   Lactate *cannot* pass electrons exergonically to NAD^+ under standard conditions because the pyruvate/lactate redox pair has a less negative E_0' ($-0.19V$) than does the $NAD^+/NADH$ redox pair ($-0.32V$), which means that the reduced form of the $NAD^+/NADH$ redox pair (i.e., NADH) will spontaneously reduce the oxidized form of the pyruvate/lactate redox pair (i.e., pyruvate). Thus the reaction will be spontaneous in the direction of pyruvate reduction to lactate, not lactate oxidation to pyruvate.

The $\Delta E_0'$ value for the reduction of lactate to pyruvate by NAD^+ is calculated as:

$$E_0' = E_0', \text{ acceptor} - E_0', \text{ donor} = -0.32V - (-0.19V) = -0.32 + 0.19 = \textbf{-0.13}V.$$

The $\Delta E_0'$ value is negative, so the transfer of electrons from lactate to NAD^+ *is not* thermodynamically spontaneous under standard conditions, as predicted above.

The $\Delta G^{o'}$ value for the reduction of NAD^+ by isocitrate is calculated as:

$$\Delta G^{o'} = -nF\Delta E_0' = -1(23{,}062)(-0.13) = +2998^{\text{cal/mol}} = \textbf{+3.00 kcal/mol.}$$

These calculations are relevant to aerobic energy metabolism because in at least some types of cells (muscle cells, for example) lactate accumulates during periods of hypoxia (low oxygen concentration) and must be reoxidized to pyruvate when the oxygen concentration rises again. The above calculations tell us that this reoxidation would not be possible under standard conditions, but under cellular conditions, the concentration ratios of lactate to pyruvate and of NAD^+ to NADH must be sufficiently high to render the reaction exergonic.

(e)   By the same reasoning as in part (d), succinate *cannot* pass electrons exergonically to NAD^+ under standard conditions because the fumarate/succinate redox pair has a much less negative E_0' ($-0.03V$) than does the $NAD^+/NADH$ redox pair ($-0.32V$), which means that the reduced form of the $NAD^+/NADH$ redox pair (i.e., NADH) will spontaneously reduce the oxidized form of the fumarate/succinate redox pair (i.e., fumarate). Thus the reaction will be spontaneous in the direction of fumarate reduction to succinate, not succinate oxidation to fumarate. The relevant calculations for $\Delta E_0'$ and $\Delta G^{o'}$ for succinate oxidation by NAD^+ are:

$$\Delta E_0' = E_0', \text{ acceptor} - E_0', \text{ donor} = -0.32V - (-0.03V) = -0.32 + 0.03 = \textbf{-0.29V.}$$

$$\Delta G^{\circ\prime} = -n\text{F}\Delta E_0' = -1(23,062)(-0.29) = +6668^{\text{cal/mol}} = \textbf{+6.67 kcal/mol.}$$

The highly positive value for $\Delta G^{\circ\prime}$ makes it virtually certain that succinate oxidation with NAD^+ as the electron acceptor is impossible under cellular conditions. It is highly unlikely that the concentration ratios of the products and reactants could overcome so highly positive at $\Delta G^{\circ\prime}$ value. Thus, NAD^+ is *not* likely to serve as the electron acceptor for the succinate dehydrogenase reaction of the TCA cycle.

(f)   By the same reasoning as in part (a), succinate *is* able to pass electrons exergonically to coenzyme Q under standard conditions because the fumarate/succinate redox pair has a more negative E_0' ($-0.03V$) than does the CoQ/CQH_2 redox pair ($+0.04V$). Thus the reaction will be spontaneous in the direction of succinate oxidation under standard conditions. The relevant calculations for E_0' and $\Delta G^{\circ\prime}$ for succinate oxidation by CoQ are:

$$E_0' = E_0', \text{ acceptor} - E_0', \text{ donor} = +0.04V - (-0.03V) = +0.04 + 0.03 = \textbf{+0.07V.}$$

$$\Delta G^{\circ\prime} = -n\text{F}\Delta E_0' = -1(23,062)(+0.07) = -1614^{\text{cal/mol}} = \textbf{-1.61 kcal/mol.}$$

These calculations mean that the transfer of electrons from succinate to CoQ will be thermodynamically spontaneous under standard conditions. It makes sense to regard coenzyme Q as the electron acceptor even though FAD is shown as the immediate electron acceptor in Figure 14-8 because the FAD involved here is tightly bound to the succinate dehydrogenase complex and passes its electrons to coenzyme Q, making the latter the eventual electron acceptor for this reaction.

14-7.   (a)   For calculation of the maximum ATP yield for an aerobic prokaryote, see Table S14-1 (but ignore numbers in parentheses; they are the values for the eukaryotic cell described in part b). The maximum ATP yield is obtained by summing the bottom line across the table: ATP yield = 8 + 6 + 24 = 38 ATP/glucose.

Table S14-1   Calculation of Maximum ATP Yield from Aerobic Oxidation of Glucose

Stage of Respiration	Glycolysis (glucose → 2 pyruvate)	Pyruvate Oxidation (2 pyruvate → 2 acetyl CoA)	TCA Cycle (2 turns)
Yield of CO_2	0	2	4
Yield of NADH	2	2	6
ATP per NADH	3 (2)	3	3
Yield of $FADH_2$	0	0	2
ATP per $FADH_2$			2
ATP from substrate-level phosphorylation	2	0	2
ATP from oxidative phosphorylation	6 (4)	6	22
MAXIMUM ATP YIELD	8 (6)	6	24

(b) For a eukaryotic cell that uses the glycerol phosphate shuttle, substitute the parenthetical values shown in the "Glycolysis" column; because the glycerol phosphate shuttle transfers electrons from NADH in the cytosol to FAD in the mitochondrion, the ATP yield per cytoplasmically generated ATP is 2 instead of 3 and the ATP yield is therefore decreased by 2: ATP yield = 6 + 6 + 24 = 36 ATP/glucose.

14-8. (a) High levels of ADP mean low levels of ATP, so it is to the advantage of the cell to activate the metabolic pathway responsible for coenzyme reduction, which can in turn give rise to ATP synthesis by electron transport.

(b) High NADH levels mean adequate reduced coenzyme for the generation of more ATP, so it makes sense to shut down the catabolic machinery of the cell.

(c) High ATP levels indicate adequate energy supply, so it makes sense that the enzyme responsible for providing the TCA cycle with more acetyl CoA substrate is shut down.

(d) High citrate levels are indicative of a sufficient supply of acetyl CoA, so it is reasonable that the key regulatory enzyme of the pathway leading to pyruvate and acetyl CoA is decreased in activity.

(e) High levels of NADH mean adequate reduced coenzyme for the generation of ATP, so it makes sense to convert PDH to the inactive (phosphorylated) form, which is what PDH kinase does.

(f) High levels of succinyl CoA signal adequate levels of TCA-cycle intermediates, so it seems reasonable to shut down further TCA cycle activity.

14-9. (a) Fluorocitrate has been characterized as the actual poison in the tissues of the animal, and one of the most pronounced of its effects is a build-up of at least one of the TCA cycle intermediates.

(b) The suspected blockage point is the conversion of citrate to isocitrate (inhibition of the enzyme aconitase), because (1) it is specifically citrate that accumulates in the tissues of poisoned animals and (2) fluorocitrate is an analogue of citrate and might be expected to bind to, and thereby to block the active site of the enzyme that metabolizes citrate in the TCA cycle.

(c) Fluoroacetate is probably activated to fluoroacetyl CoA and condensed onto oxaloacetate by citrate synthase to generate fluorocitrate.

(d) The ingested compound is itself not toxic, but is converted into a lethal metabolite in vivo (that is, a lethal compound is synthesized in vivo from a nonlethal precursor).

14-10. (a) $DHAP + NADH_{cytosol} + H^+ \xrightarrow{\text{in the cytosol}} Gly\text{-}3\text{-}P + NAD^+_{cytosol}$

$FAD_{mitochondrion} + Gly\text{-}3\text{-}P \xrightarrow{\text{in the inner membrane}} DHAP + FADH_{2,\ mitochondrion}$

(b) $NADH_{cytosol} + H^+ + FAD_{mitochondrion} \rightarrow NAD^+_{cytosol} + FADH_{2,\ mitochondrion}$

To calculate the difference in standard reduction potentials, use equation 14-9 and appropriate data from Table 14-2 in the textbook.

$\Delta E_0' = \Delta E_0'_{,\ acceptor} - \Delta E_0'_{,\ donor} = -0.18 - (-0.32) = +0.14$ **V**

To calculate the corresponding standard free-energy change, use equation 14-11 in the textbook:

$\Delta G^{\circ\prime} = -nF\Delta E_0' = -2(23{,}062)\ (+0.14) = -6460$ cal/mol = **-6.46 kcal/mol.**

The inward movement of electrons is feasible, given the highly negative $\Delta G^{\circ\prime}$ value.

(c) $FADH_{2,\ mitochondrion} + COQ_{mitochondrion} \rightarrow FAD_{mitochondrion} + CoQH_{2,\ mitochondrion}$

$\Delta E_0' = \Delta E_0'_{,\ acceptor} - \Delta E_0'_{,\ donor} = +0.04 - (-0.18) = +0.22$ **V**

$\Delta G^{\circ\prime} = -nF\Delta E_0' = -2(23{,}062)\ (+0.22) = -10{,}150$ cal/mol = **-10.1 kcal/mol.**

This transfer is also thermodynamically feasible under standard conditions.

(d) $NADH_{cytosol} + H^+ + CoQ_{mitochondrion} \longrightarrow NAD^+_{cytosol} + CoQH_{2,\ mitochondrion}$

Because this reaction is the sum of the reactions in parts b and c, the values for $\Delta E_0'$ and $\Delta G^{\circ\prime}$ can be obtained as the sums of the values for the two component reactions:

$\Delta E_0' = +0.14 + 0.22$ V = **+0.36 V**

$\Delta G^{\circ\prime} = 6{,}460 - 10{,}150 = -16{,}600$ cal/mol = **-16.1 kcal/mol.**

Given that both component reactions are exergonic under standard conditions, it is not surprising that the overall reaction is exergonic—*highly* exergonic, in fact.

(e) $\Delta G' = \Delta G^{\circ\prime} + RT \ln ([NAD^+]\ [CoQH_2]/[NADH]\ [CoQ])$

$= -16{,}600 + (1.987)\ (25 + 273) \ln (2.0/5.0)$

$= -16{,}600 + 592 \ln (0.4) = -16{,}600 + (592)\ (-0.916)$

$= -16{,}600 - 542 = -16{,}060$ cal/mol = **-16.1 kcal/mol.**

(f)   No, it is not affected by the concentration of the intermediate in the process because free energy is a thermodynamic parameter and is therefore a measure of the difference in energy content between the reactants and products only.

14-11. (a)   Given that ATP synthesis does not occur, the energy is lost as heat. For a newborn baby, the heat generated in this way may be critical to maintenance of body temperature.

(b)   One would expect to find more thermogenin in a hiberating bear because the need for additional heat is clearly more critical during hibernation, when the external temperature is likely to be colder and bodily activity much less than in the case of a physically active bear.

(c)   The localization of thermogenin to the inner mitochondrial membrane and its mode of action as an uncoupler of electron transport make it likely that thermogenin is a proton translocator that allows electrons to move exergonically into the matrix of the mitochondrion just as F_o does, but without concomitant ATP generation. To test this hypothesis, one could prepare vesicles consisting of phospholipid bilayers with and without thermogenin. The vesicles could be prepared in a mildly alkaline solution (pH 8.0, for example), then transferred to an acidic solution (pH 5.0, for example). If thermogenin is a proton translocator, protons should enter the vesicles containing this protein, such that the pH within the vesicle should quickly equilibrate with the external pH (i.e., pH 5) whereas the pH within the vesicles prepared without thermogenin should remain at the original pH (i.e., pH 8).

14-12. (a)   Every molecule of glucose generates 2 molecules of pyruvate and 2 molecules of cytosolic NADH, with the latter giving rise to 2 molecules of mitochondrial $FADH_2$. In the mitochondrion, the complete oxidation of 1 molecule of pyruvate to carbon dioxide and water generates 1 molecule of $FADH_2$ and 4 molecules of NADH. Thus, the yield of mitochondrial coenzymes per molecule of glucose is 8 molecules of NADH and 2 molecules of $FADH_2$. The reoxidation of 8 molecules of NADH means that 8 electron pairs must traverse respiratory complexes I, III, and IV, each of which couples electron transport to proton pumping. Electron transport from NADH will therefore result in the outward pumping of 72 ($8 \times 3 \times 3$) protons. Similarly, the reoxidation of 4 molecules of $FADH_2$ means that 4 electron pairs must traverse respiratory complexes II, III, and IV, only two of which couple electron transport to proton pumping. Electron transport from $FADH_2$ will therefore result in the outward pumping of 24 ($4 \times 2 \times 3$) protons. The total number of protons pumped outward per molecule of glucose oxidized is therefore 96 ($72 + 24$). At 3 protons per ATP, 96 protons could generate 32 molecules of ATP if all the energy of the proton gradient were used for this and no other purpose.

(b)   One molecule of glucose generates two molecules of pyruvate, which require a total of 4 protons for transport into the mitochondrion, leaving a net yield of **92 protons per molecule of glucose.**

(c)   To generate one molecule of ATP requires four protons—the three that must pass inward through the F_oF_1 complex and the additional proton for inward transport of the needed phosphate ion. The 92 protons from part b can therefore be used to generate **23 ATP molecules.**

(d)   Taking into account the two ATP molecules generated by glycolysis and the two generated by the TCA cycle, the total yield under these conditions is **27 molecules of ATP per molecule of glucose** and the overall equation is

$$C_6H_{12}O_6 + 6\ O_2 + 27\ ADP + 27\ P_i \longrightarrow 6\ CO_2 + 27\ ATP + 33\ H_2O.$$

14-13. (a)   The products of fatty acid oxidation and their fate are:

Acetyl CoA: converted via succinate to hexoses.

$FADH_2$: reoxidized to FAD by oxidase activity, consuming oxygen and generating hydrogen peroxide.

NADH: cannot be reoxidized within glyoxysome; must be passed to outside (see part f below).

(b)   One molecule of palmitate gives rise to eight molecules of acetyl CoA and hence to four molecules of succinate. This requires seven cycles of β oxidation, so seven molecules each of $FADH_2$ and NADH are formed.

(c)   Four molecules of succinate give rise to four molecules of phosphoenolpyruvate and hence to two hexose molecules. The sequence from succinate to oxalo-acetate is part of the TCA cycle and occurs in the mitochondrion; the remainder of the pathway takes place in the cytosol. Other products include the $FADH_2$ and NADH formed by mitochondrial oxidation of succinate and malate, respectively; the carbon dioxide produced in the conversion of oxaloacetate to phosphoenolpyruvate; and the NAD^+ generated as glycerate-3-phosphate is reduced to glyceraldehyde-3-phosphate in the reversal of the glycolytic pathway necessary to convert phosphoenolpyruvate to hexoses.

(d)   Because a molecule of palmitate gives rise to two hexose molecules, the end product will be a single sucrose molecule. Of the original 16 carbon atoms in palmitate, 12 appear in sucrose. The other 4 carbon atoms are evolved as carbon dioxide when oxaloacetate is converted to phosphoenolpyruvate.

(e)   Because one molecule of palmitate ($C_{16}H_{32}O_2$; MW = 256) gives rise to one molecule of sucrose ($C_{12}H_{22}O_{11}$; MW = 342), each gram of stored palmitate will yield about 1.34 (342/256) g of sucrose.

(f)   NADH moves outward for reoxidation outside the glyoxysome, NAD^+ moves inward to allow continued oxidation of fat, and oxygen moves inward to provide for direct oxidation of $FADH_2$. The ATP needed for activation of fatty acids (reaction FA-1) must be supplied from outside the organelle, so ATP moves inward also, presumably accompanied by exchange of ADP outward.

14-14. (a)   The electron donor in this system is β-hydroxybutyrate, which can oxidized to β-ketobutyrate and the electron acceptor is oxidized cytochrome *c*, which can be reduced by taking up a single electron. The most likely pathway of electron transport is from β-hydroxybutyrate via NAD^+, complex I and complex III to cytochrome *c*.

(b)   Assuming the pathway specified in part a, the ATP yield is likely to be 2 moles of ATP per mole of β-hydroxybutyrate oxidized, one each by complexes I and III.

The reaction in this system is as follows:

β-hydroxybutyrate + 2 cyt $c_{oxidized}$ + 2ADP + 2 P_i →

β-ketobutyrate + 2 cyt $c_{reduced}$ + $2H^+$ + 2ATP + $2H_2O$

(c)   Cyanide is a potent inhibitor of cytochrome *c* oxidase. It is present to ensure that the electrons picked up by cytochrome *c* are not passed on down the system to O_2, which is the acceptor when the whole electron transport system is operative. In the absence of cyanide, the electrons from β-hydroxybutyrate would be transferred all the way to oxygen, which would defeat the purpose of the experiment.

(d)   The TCA cycle would not be active because no substrate for the TCA cycle (i.e., no pyruvate, acetyl CoA, or fatty acid) is present in the assay system.

(e)   It is important that the β-ketobutyrate generated by oxidation of β-hydroxybutyrate cannot be further metabolized, lest other oxidative reactions occur that would generate alternative substrates for the electron transport system. If lactate had been used instead—and if isolated mitochondria can oxidize lactate as they can oxidize β-hydroxybutyrate—pyruvate would be formed, which would activate the TCA cycle, thereby generating additional substrates for the electron transport system (NADH and FADH$_2$) and rendering the experiment useless for its intended purpose.

15

Phototropic Energy Metabolism: Photosynthesis

15-1. (a)   T   (c)   F   (e)   F
        (b)   F   (d)   F   (f)   T

15-2. (a)   Phosphoenolpyruvate carboxylase (PEPCase), the enzyme that catalyzes initial carbon dioxide fixation in a C_4 plant, has a much higher affinity for carbon dioxide than does rubisco, the enzyme that catalyzes the initial carbon dioxide fixation in C_3 plants. Furthermore, PEPCase lacks the oxygenase activity displayed by rubisco. In C_4 plants, carbon dioxide fixed by PEPCase in mesophyll cells is released in bundle sheath cells, where the Calvin cycle occurs. The enhanced concentration of carbon dioxide in the bundle sheath cells favors rubisco's carboxylase activity and minimizes its oxygenase activity.

        (b)   PEPCase has a much higher affinity for carbon dioxide than does rubisco. Thus, the lower the ambient carbon dioxide concentration, the greater the competitive advantage of a plant species in which the Hatch-Slack cycle precedes the Calvin cycle and thereby increases the concentration of carbon dioxide in the vicinity of rubisco.

        (c)   As both plants fix carbon dioxide, the concentration of carbon dioxide within the sealed container will decrease, giving the C_4 plant a greater and greater competitive advantage; see (b). Eventually, the carbon dioxide concentration will become so low that rubisco of the C_3 plant will not function effectively at all, while the Hatch-Slack cycle of the C_4 plant continues to concentrate carbon dioxide in bundle sheath cells. Beyond this point, the C_3 plant will release more carbon dioxide through mitochondrial respiration than it acquires through photosynthesis, resulting in a net loss of reduced carbon and decline in food reserves. The C_4 plant continues to fix the released carbon dioxide, gaining at the expense of the C_3 plant until the C_3 plant exhausts its reserves and perishes.

15-3. (1)   ATP is highly polar and would be difficult to move across membranes.

        (2)   ATP would be a very inefficient way to move energy about, because it has a molecular weight of about 500 daltons and contains only two high-energy phosphate bonds. Sucrose, in contrast, has a molecular weight of about 340 daltons and can give rise to 72 molecules of ATP elsewhere in the plant.

        (3)   Generation of ATP stops as soon as the sun sets and there is no practical way to store enough of it to meet the ongoing energy needs of the organism at night.

15-4.   (a)   PGA down, RuBP up. With little or no carbon dioxide on hand to generate PGA from RuBP, the former will decrease in concentration and the latter will accumulate.

(b)   PGA up, RuBP down. In the presence of green light only, the light-dependent reactions will function minimally if at all, which means that there will be little or no ATP and NADPH to drive the reduction of PGA to glyceraldehyde-3-phosphate from which RuBP is regenerated. As a result, PGA will accumulate and RuBP will be depleted.

(c)   PGA up, RuBP down. The same reasoning as in (b); the products of the light-dependent reactions are required to keep the Calvin cycle functioning, and an inhibitor of Photosystem II will effectively prevent the formation of NADPH and will almost certainly reduce the formation of ATP as well.

(d)   PGA up, RuBP unchanged. Decreased oxygen availability will lessen oxygenase activity of rubisco, thereby allowing more net carboxylation and hence more PGA formation. RuBP consumption, on the other hand, is not likely to change; it is simply being used more for carboxylation and less for oxygenation.

13-5.   (a)   The reaction is:

$$\text{Light} + CO_2 + 2\ H_2S \longrightarrow [CH_2O] + 2\ S + H_2O$$

Or, with the stoichiometry required for the formation of one triose molecule:

$$\text{Light} + 3\ CO_2 + 6\ H_2S \longrightarrow C_3H_6O_3 + 6\ S + 3\ H2O$$

(b)   The S/H_2S couple has a standard reduction potential of -0.27 V and may donate electrons either to the bacteriochlorophyll molecules at the reaction center of a bacterial photosystem or to a cytochrome. Light, either directly or indirectly, can then drive electrons energetically uphill, thereby reducing NAD^+ to NADH and providing reductant for carbon assimilation. The O_2/H_2O couple does not have a negative enough standard reduction potential to enable it to donate electrons to this system.

(c)   The negative standard reduction potential of the S/H_2S couple makes H_2S a suitable electron donor under low light; only one low-energy photon is required to excite an electron obtained from H_2S to a level sufficient for reducing $NADP^+$ to NADPH. Furthermore, organisms that are adapted to anaerobic environments are often susceptible to harm by the O_2 generated when water is used as an electron donor.

(d)   No, the standard reduction potential of the Fe^{3+}/Fe^{2+} couple is much too positive to allow spontaneous transfer of electrons from ferrous ions to bacteriochlorophyll or cytochrome.

15-6.   The flow of energy is as follows:

photon → excited electron in accessory pigment → excited electron in chlorophyll molecule → excited electron in special chlorophyll molecule at the reaction center of a photosystem → reduced organic electron acceptor → reduced intermediates along the electron transport system → electrochemical proton gradient across the thylakoid membrane → ATP → high-energy phosphate bond of glycerate-1,3-bisphosphate → glyceraldehyde-3-phosphate → fructose-6-phosphate → intermediate monosaccharides → starch.

15-7.   Although both the mint sprig and the mouse require air, they do not depend on it for the same reason. The mint sprig needs carbon dioxide, while the mouse needs oxygen. In fact, the photosynthetic activity of the mint sprig enhances the supply of oxygen in the air, rendering it "not at all inconvenient" to the mouse. The observation was important in establishing that whatever the photosynthetic organism acquired from air, it was obviously not the same material that a respiring animal depended on.

15-8.   (a)   Valid   (b)   Valid   (c)   Valid

15-9.   (a)   Lower; more demand for NADPH via noncyclic electron flow.

(b)   Higher; more demand for ATP via cyclic electron flow.

(c)   Higher in both mesophyll cells and bundle sheath cells; regeneration of phosphoenol pyruvate in mesophyll cells requires ATP, while transfer of reducing equivalents from mesophyll cells to bundle sheath cells lowers the demand for NADPH.

(d)   Higher; noncyclic electron flow is inhibited.

(e)   Lower; cyclic electron flow is inhibited and noncyclic electron flow is enhanced by the presence of an artificial electron acceptor that acts as an electron sink.

(f)   Lower; for every three carbon atoms salvaged, a molecule of ammonia must be reassimilated at the expense of one ATP and two NADPH, and a glycerate molecule must be phosphorylated at the expense of one ATP.

15-10. (1)   Not all light will be of an appropriate wavelength (i.e., green light is reflected).

(2)   Many leaves may be shaded by leaves in the canopy above them.

(3)   The difference in energy between photons absorbed by light-harvesting pigments and the energy that actually excites an electron within the reaction center of a photosystem is lost as heat and entropy.

(4)   The capacity of a photosystem to absorb energy may be saturated by a level of light far below the intensity of sunlight.

15-11. (a)   Stroma.

(b)   Through photosystem I and the cytochrome b_6/f complex.

(c)   Stroma.

(d)   Nonappressed regions of the thylakoid membrane.

(e)   Lipid phase of the thylakoid membrane.

(f)   Across the thylakoid membrane from the stroma into the lumen.

(g)   Photosystem I.

(h)   Stroma.

(i)   Photosystems and light-harvesting complexes.

(j)   Photosystem II.

15-12. (a)   Across the outer and inner chloroplast membranes from the cytosol to the stroma.

(b)   Across the outer and inner chloroplast membranes from the cytosol to the stroma.

(c)   No.

(d)   No.

(e)   Across the inner and outer chloroplast membranes from the stroma to the cytosol.

(f)   No.

(g)   No.

(h)   Across the thylakoid membrane from the lumen to the stroma, and across the inner and outer chloroplast membranes from the stroma to the cytosol.

(i)   Across the thylakoid membrane from the stroma to the lumen.

(j)   In the bundle sheath cells of C_4 plants: across the inner and outer chloroplast membranes from the stroma to the cytosol.

In the mesophyll cells of C_4 plants: across the outer and inner chloroplast membranes from the cytosol to the stroma.

15-13. During the night: CO_2 in atmosphere → stomata → mesophyll cell → bicarbonate → oxaloacetate (in cytosol) → malate (in cytosol) → malate (in vacuole).
During the day: malate (in vacuole) → malate (in cytosol) → carbon dioxide (in cytosol) → Calvin cycle → glyceraldehyde-3-phosphate.
CAM plants are able to fix CO_2 and accumulate carbon in the form of organic acids at night, when the temperature is likely to be cooler and the loss of water through open stomata thereby minimized. Open stomata are necessary for gas exchange between the interior and exterior of the leaf. During the day, when solar energy is available but the potential for water loss is highest, CAM plants close their stomata and channel carbon from malate into the Calvin cycle. C_3 and C_4 plants, on the other hand, must open their stomata during the day to acquire sufficient carbon dioxide.

15-14. (a)   The inner membrane systems found in mitochondria and chloroplasts may be derived from ingested bacteria, while the outer membrane systems may be derived from nucleated hosts. An identical arrangement of membranes follows phagocytosis: the ingested bacterium is surrounded by the membrane of the phagocytic vacuole,. The thylakoid membrane systems found in chloroplasts may also be derived from ingested bacteria, because thylakoids form from invaginations of the inner membrane that pinch off to generate a separate membrane system.

(b)   One of the most important features the bacteria might have dispensed with is the cell wall. Without cell walls, the bacteria might burst due to osmotic pressure or might be more vulnerable to invasion by viruses.

(c)   See part b.

(d)   Features that mitochondria have retained while peroxisomes have not include (i) DNA replication, (ii) RNA synthesis, (iii) protein synthesis, (iv) an inner as well as an outer membrane, (v) electron transport, and (vi) ATP synthesis.

(e)   Peroxisomes might have enabled ancient cells to (i) expand the use of oxidases—which generate the potentially toxic compound H_2O_2—for a variety of metabolic processes; (ii) generate acetyl coenzyme A from long-chain fatty acids, providing additional carbon sources for the cell; and (iii) reduce the toxicity of a variety of compounds by using the molecules as electron donors for breaking down H_2O_2.

16

The Structural Basis of Cellular Information: DNA, Chromosomes, and the Nucleus

16-1. (a)   I   (c)   I   (e)   N   (g)   B

      (b)   B   (d)   B   (f)   I   (h)   N

16-2. (a)   It made their experiment possible, because it allowed them to assay various fractions of S cells for transforming activity in culture.

      (b)   This was the very hypothesis their experiment was designed to test; all that was left to do was to devise a means of identifying the nucleic acid if it did indeed "flow into the cell."

      (c)   The concept of base pairing by hydrogen bonding between pyrimidines and purines on opposite strands turned out to be a vital clue to the double-stranded structure of the DNA molecule.

      (d)   By showing that a virus was indeed like a "little hypodermic needle" capable of injecting its nucleic acid into the cell, Hershey and Chase were able to explain Anderson and Herriott's observation in terms of an osmotic shock that causes the viruses to empty their nucleic acid contents into the medium.

16-3. (a)   T   (c)   F   (e)   F

      (b)   T   (d)   T

16-4. Figure S16-1 depicts the phage DNA, which has a total of 10.5 kilobase pairs. The vertical arrows indicate sites on the DNA that are cut by restriction enzymes X and Y.

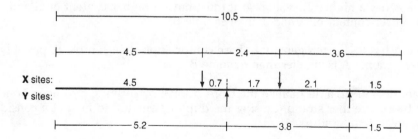

Figure S16-1 Restriction Mapping. Shown here is the restriction map of the bacteriophage DNA, with vertical arrows to indicate sites on the DNA that are cleaved by restriction enzymes X and Y. See Problem 16-4.

16-5. The DNA fragment to be analyzed is illustrated in Figure S16-2a, with the four-base primer region shown on the left and the unknown sequence on the right. The base sequence of the DNA fragment is determined by mixing together: 1) a single-stranded (denatured) preparation of the DNA fragment, 2) the deoxynucleotides dATP, dCTP, dTTP, and dGTP, 3) the dideoxynucleotides ddATP, ddCTP, ddTTP, and ddGTP, each labeled with a fluorescent dye of a different color, e.g., ddATP = red, ddCTP = blue, ddTTP = orange, and ddGTP = green, 4) DNA polymerase, and 5) the single-stranded primer. After incubation, the reaction products are separated by gel electrophoresis. As indicated in Figure S16-2b, the pattern of colored bands in the gel will reveal the sequence of the original DNA.

(a)

$$3'-A-G-C-G-C-T-A-T-A-G-C-G-C-T-5'$$
$$5'-T-C-G-C-G-A-T-A-T-C-G-C-G-A-3'$$

Primer   Unknown sequence

(b)

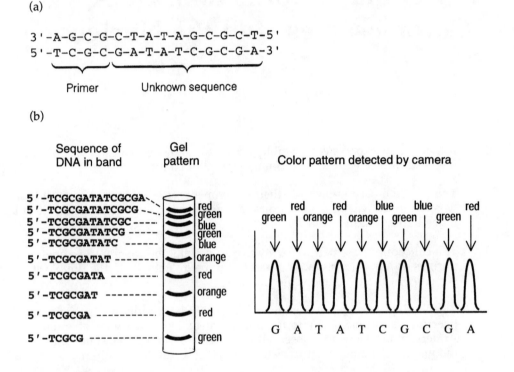

Figure S16-2 **DNA Sequencing.** (a) The DNA fragment to be analyzed. (b) The pattern of bands obtained using dye-labeled dideoxynucleotides, showing the base sequence of the DNA in each band and the pattern of colors detected by the camera. See Problem 16-5.

16-6. (a) Sample A has a higher T_m value, so it must have a higher content of G and C and a lower content of A and T than sample B.

(b) Sample A has less heterogeneity in base composition than sample B, possibly because genome A is smaller than genome B.

(c) Formamide and urea destabilize the DNA duplex by forming hydrogen bonds to the bases of either strand, causing the duplex structure to melt at a considerably lower temperature.

16-7.   (a)   If the two DNA samples had nucleotide sequences that were similar to each other, but not exactly the same, strands from one sample could hybridize to strands from the other sample. Because the resulting hybrids would not be exactly complementary, you would get a lower melting temperature because DNA molecules in which the two strands of the double helix are properly base-paired at each position melt at higher temperatures than DNA in which the two strands are not perfectly complementary.

         (b)   To test your hypothesis, you might carry out base sequencing analysis of the two DNA samples. If your hypothesis were correct, the two DNA base sequences should be similar, but not exactly the same.

         (c)   There are two possible interpretations of such a result. First, the two initial samples of DNA might have been identical, in which case mixing together their individual strands would have no effect on the melting temperature following reassociation. Alternatively, the two initial samples may have been totally unrelated to each other in nucleotide sequence. In this case, strands from one sample could not hybridize with strands from the other sample during the reassociation step. Thus, the strands from sample "A" would only hybridize with complementary strands from sample "A" during the reassociation step, and strands from sample "B" would only hybridize with complementary strands from sample "B". Hence the original melting temperature of each sample (92°C) would be retained after reassociation. The fact that both samples initially had the same melting temperature does not mean that they were related in sequence, although it does indicate that the relative proportion of GC base pairs in both samples was similar.

16-8.   (a)   The fact that the DNA fragments are all multiples of a basic unit 260 bp in length suggests that proteins are clustered along the DNA in a regular pattern that repeats at intervals of roughly 260 bp. Such a regular distribution of protein clusters suggests the existence of nucleosomes, even though the distance between them appears to be longer than the more typical value of 200 bp described in the chapter.

         (b)   In this case, each nucleosome appears to be associated with 260 bp of DNA.

         (c)   You would expect to see a series of fractions containing differing numbers of particles. The smallest fraction would contain single particles, the next smallest fraction would contain clusters of two particles, the succeeding fraction would contain clusters of three particles, and so forth (see Figure 16-19 in the textbook). If the DNA were isolated from these fractions and analyzed by gel electrophoresis, the DNA from the fraction containing single particles would be expected to measure 260 bp in length, the DNA from the fraction containing clusters of two particles would be expected to measure 520 bp in length, the DNA from the fraction containing clusters of three particles would be expected to measure 780 bp in length, and so on.

         (d)   This experimental observation indicates that the nucleosomal core particle contains 146 bp of DNA. Since a total of 260 bp of DNA is associated with the nucleosome, the linker must be 260 − 146 = 114 bp.

16-9.   (a)   Unlike most membranes, the nuclear envelope appears to be freely permeable to a polar organic molecule.

(b)   The aqueous channels in nuclear pore complexes have diameters of at least 5.5 nm, but not as great as 15 nm.

(c)   The stained pores contain complexes of RNA and protein, probably ribonucleoprotein particles caught in transit.

(d)   The NLS only triggers transport from cytoplasm to nucleus, not from nucleus to cytoplasm, suggesting that NLS receptor proteins (importins) function only in the cytoplasm.

(e)   The nuclear membranes and the endoplasmic reticulum are likely to have a common origin.

(f)   Ribosomal proteins must pass inward from the cytoplasm to the nucleus at a rate adequate to sustain ribosomal subunit assembly; ribosomal subunits must move outward from the nucleus to the cytoplasm at a rate commensurate with their rate of assembly.

(g)   Nucleoli are probably responsible for rRNA synthesis.

(h)   The integrity of the nucleus does not seem to depend entirely on the envelope.

16-10. (a)   F; nucleoli are structures made of DNA, RNA, and protein that are present in the eukaryotic nucleus.

(b)   T

(c)   F; the DNA of nucleoli carries the cell's rRNA genes, which are present in cluster of multiple copies.

(d)   F; a single nucleolus may contain a number of NORs.

(e)   T

(f)   T

CHAPTER

17

The Cell Cycle: DNA Replication, Mitosis, and Cancer

17-1. (a) S   (d) G1   (g) G1, G2, S   (j) G1, G2, M

(b) M   (e) M   (h) M

(c) M   (f) G1, G2, S   (i) G1, S, G2, M

17-2. (a) Mitotic index = (30 + 20 + 20 + 10 + 20)/1000 = 0.1 = **10%**.

(b) 3% of time in prophase, 2% in prometaphase, 2% in metaphase, 1% in anaphase, 2% in telophase, 40% in G1, 30% in S, and 20% in G2.

(c) Because G2 is 20% of the cell cycle and lasts 4 hours, the whole cycle must be 4/0.2 = **20 hours.**

(d) 0.6 h in prophase; 0.4 h each in prometaphase, metaphase, and telophase; 0.2 h in anaphase; 8 h in G1; 6 h in S; and 4 h in G2.

(e) The first appearance of label in prophase nuclei would have to be observed.

(f) About 30%, because all cells in S phase will incorporate label immediately, and on the average about 30% of the cells in the culture should be in S phase at any one time.

17-3. (a) Figure S17-1a illustrates the expected progeny molecules for two rounds of DNA replication according to the conservative model (on left) and the dispersive model (on right)

(b) Figure S17-1b depicts the distribution of DNA bands in cesium chloride gradients after two rounds of DNA replication according to the conservative model (on left) and the dispersive model (on right).

(a)

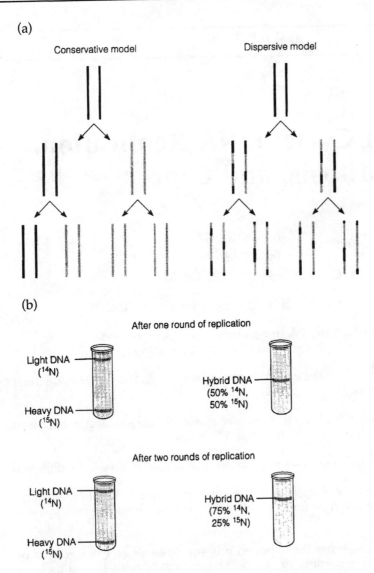

(b)

Figure S17-1   Meselson and Stahl Revisited. (a) The progeny molecules for two rounds of DNA replication according to the conservative (left) and dispersive (right) alternatives to semi-conservative replication. (b) The distribution of DNA bands in cesium chloride gradients after one and two rounds of replication by the conservative (left) and dispersive (right) alternatives to semi-conservative replication. See Problem 17-3. (It is precisely because Meselson and Stahl did not observe either of these patterns that these models were discarded in favor of the semiconservative model depicted in Figure 17-4 of the textbook.)

17-4.   Your sketch should look something like step 7 of Figure 17-13 in the textbook, with single-stranded DNA identified as the template, Okazaki fragments identified as newly forming pieces of DNA on the lagging strand, short segments of RNA on the Okazaki fragments identified as primers, and molecules of single-strand binding protein bound to the single-stranded DNA near the point at which the fork is opening. In addition, your sketch should include the following enzymes:

Helicase: unwinds double-stranded DNA.

Gyrase: nicks DNA ahead of the replication fork to relax the supercoiling induced in the DNA by helicase activity.

Primase: synthesizes short segments of RNA used as primer for DNA synthesis.

DNA polymerase III elongates the growing segments of DNA.

DNA polymerase I: removes RNA primer and replaces it with deoxyribonucleotides.

Ligase: links discontinuous fragments of DNA covalently into a continuous strand.

17-5. (a) R; in *E. coli*, exonucleolytic activity is needed for RNA primer removal and for proofreading.

(b) R; in *E. coli*, synthesis is continuous on one strand and discontinuous (with Okazaki fragments) on the other.

(c) NB; the degree of sequence reiteration is not likely to have any effect on the mode of replication of the DNA.

(d) S; Okazaki fragments have short RNA (ribose) segments as well as longer DNA (deoxyribose) segments.

(e) R; this finding is not consistent with the semiconservative mode of replication found for *E. coli* DNA.

17-6. Figure S17-2 depicts one possible solution, which depends on the parental DNA molecules having symmetric base sequences at the ends as shown. This mechanism requires two enzymes: DNA polymerase and a second enzyme that cuts the looped TAGGTA structure at its 5' end. Note that there are numerous other possible solutions as well.

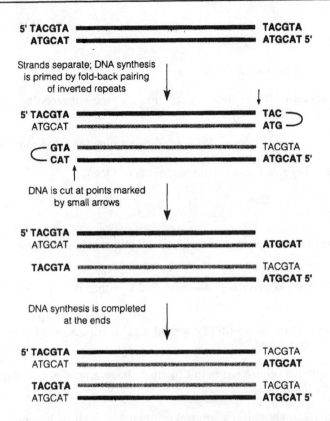

Figure S17-2   Possible Primitive Process of DNA Replication. Shown here is one possible model for DNA synthesis that would allow complete daughter molecules of DNA to be produced by a leading-strand replication mechanism. See Problem 17-6.

17-7.   (a)   Without origins of replication, the DNA would not be replicated during S phase, and only one of the two daughter cells resulting from cell division would contain this chromosome. In the next cellular generation, only one of four progeny cells would carry this chromosome, and so on, until the chromosome was effectively diluted out from the cell population.

(b)   Without a centromere, kinetochores would not form on the duplicated chromosome during mitosis, and the two chromatids would not segregate to different daughter cells. One daughter cell would end up with an extra copy of the chromosome, and the other cell would lack a copy.

(c)   With each round of DNA replication, the progeny DNA molecules would be slightly shortened at both ends, as is usually the case. However, at the end lacking a telomere for protection, the chromosome would presumably lose important genetic information in a relatively small number of cell generations. Even more dire consequences could result, especially if the chromosome in question were in a germline cell. The telomere lengthening enzyme, telomerase, would be unable to function at one end of the DNA, and eventually the DNA (and chromosome) would be completely degraded from that end.

17-8. (a)   DP   (d)   DF   (g)   DP, DA, DF   (j)   None

      (b)   DF   (e)   None   (h)   DP, DA, DF

      (c)   DP   (f)   DA   (i)   DP, DA, DF

17-9. (a)   Uracil arises from cytosine, hypoxanthine from adenine, and xanthine from guanine.

      (b)   Thymine does not contain an amino group and therefore cannot be deaminated.

      (c)   The excision process would not be able to distinguish between a base that has been generated by deamination and a naturally occurring base.

      (d)   The base 5-methylcytosine yields thymine upon deamination, and there is no mechanism to detect and excise thymine bases. Deamination of 5-methylcytosine will therefore go unrepaired, and every such deamination will convert a C into a T.

17-10. (a)   Inconsistent. If the microtubules disassembled at the spindle pole, both the dark (photobleached) band and the chromosomes would move at the same rate. If such were the case, the chromosomes would remain the same distance from the dark band, rather than appearing to move toward it.

      (b)   Consistent. (However, the *main* explanation for chromosome movement may be the action of motor proteins. The disassembly of microtubules at the kinetochores is probably a secondary event.)

      (c)   Inconsistent. Again, if microtubules were contracting in some way, the chromosomes and the photobleached band would probably move at the same rate.

      (d)   Consistent. (This statement appears to be the main explanation for chromosome movement.)

17-11. (a)   G1 has the 2C amount of DNA; G2 has the 4C amount.

      (b)   DNA is being synthesized during S; little or no synthesis occurs during G1.

      (c)   Chromosomes are in an extended form during G2, but in a condensed form during most of M phase. A cell near the end of M phase may have decondensed chromosomes, but it will be distinguished by having two nuclei not yet completely partitioned into daughter cells.

      (d)   Chromosomes are in an extended form during G1, but in a condensed form during most of M phase. See also answer (c).

17-12. (a)   T   (c)   F   (e)   T

      (b)   F   (d)   NP

17-13. (a)   MPF (mitotic Cdk bound to mitotic cyclin) is normally activated at the end of the G2 phase of the cell cycle, when it then catalyzes the phosphorylation of lamins, histone H1, and the condensin complex. These phosphorylations in turn trigger nuclear envelope breakdown and chromosome condensation. By experimentally introducing activated MPF into cells that have just emerged from S phase, chromosome condensation and nuclear envelope breakdown can therefore be triggered prematurely.

(b) The ability of an indestructible form of cyclin to block a cell's exit from mitosis shows that the destruction of mitotic cyclin—and hence the inactivation of MPF—are required before cells can complete mitosis and enter a new cell cycle.

(c) The breakdown of the nuclear envelope at the beginning of mitosis is triggered by the phosphorylation of lamins catalyzed by MPF. The discovery that cells deficient in protein phosphatase activity have difficulty reconstructing the nuclear envelope suggests that proteins which have been phosphorylated by MPF at the beginning of mitosis must be dephosphorylated again before cells can complete mitosis.

17-14. (a) In the absence of other data, three alternative interpretations of this observation are plausible. 1) EGF might trigger the activation of the Ras protein, which in turn triggers the activation of the Raf protein. 2) EGF might trigger the activation of the Raf protein, which in turn triggers the activation of the Ras protein. 3) EGF might trigger the activation of the Ras and Raf proteins via separate pathways.

(b) Since Raf can no longer be activated as efficiently in cells lacking a functional Ras protein, this suggests that the activated Ras protein normally induces the activation of Raf.

(c) Three obvious possibilities are: 1) the Raf protein might be a mutant, hyperactive protein, 2) these cancer cells might contain a mutated, hyperactive form of some component that lies before the Raf protein in the Ras pathway (e.g., a hyperactive Ras protein or a hyperactive EGF receptor), or 3) these cancer cells might have lost a tumor suppressor gene coding for a protein that normally inhibits the activity of the Ras protein. A simple way to distinguish among these alternatives is to purify DNA from these cancer cells and see if it causes normal cells to become malignant in a DNA transfection assay. If it does, this suggests the presence of an oncogene and hence favors interpretations 1) or 2). To further distinguish between these possibilities, you could clone the oncogene and see if its base sequence is similar to the normal *raf* gene, which would favor interpretation 1).

17-15. (a) PO, TS. Proto-oncogenes and tumor suppressor genes are both found in normal cells. Oncogenes are genes whose presence can cause cancer, and so are not found in normal cells.

(b) PO, OG. Some proto-oncogenes code for normal growth factors (e.g., the proto-oncogene that codes for PDGF). Although oncogenes often code for abnormal versions of proto-oncogene proteins, they can also code for excessive quantities of a normal protein encoded by a proto-oncogene. Hence an oncogene could code for excessive amounts of a normal growth factor.

(c) OG, PO, TS. Oncogenes can obviously be found in cancer cells. However, cancer cells also contain many normal proto-oncogenes and tumor suppressor genes (only a small number of the dozens of proto-oncogenes and tumor suppressor genes in a normal cell need to be mutated to trigger the development of cancer.)

(d) OG. Oncogenes are the only genes that are uniquely present in cancer cells.

(e)   OG. Oncogenes are genes whose presence can cause cancer.

(f)   TS. Tumor suppressor genes are genes whose absence can cause cancer.

(g)   TS, PO. Tumor suppressor genes and proto-oncogenes are present in both normal cells and cancer cells. (Their presence in cancer cells is explained by the fact that only a small number of the dozens of proto-oncogenes and tumor suppressor genes present in a normal cell need to be mutated to trigger the development of cancer.)

18

Sexual Reproduction, Meiosis, and Genetic Recombination

18-1.   (a)   S

(b)   A

(c)   B. However, in sexual reproduction, a mutation in one of a homologous pair of chromosomes will likely be inherited by only some of the offspring (50%, on average).

(d)   B. However, many more such offspring will result from sexual reproduction, where their occurrence is not dependent on new mutations, which are rare.

(e)   B

18-2.   (a) $2n = 4$

(b)   Correct order: F, E, D, B, A, C.

A: metaphase II

B: telophase I (and cytokinesis)

C: anaphase II (and cytokinesis)

D: metaphase I

E: late prophase I (diplotene)

F: early prophase I (leptotene)

(c)   Between metaphase I (D) and telophase I (B).

(d)   Between leptotene (F) and diplotene (E) of prophase I.

18-3.   (a)   Homologous chromosomes pair at meiotic metaphase I, but not at mitotic metaphase.

(b)   Meiotic metaphase II has only one half as many chromosomes as does mitotic metaphase.

(c)   Compared to metaphase I, metaphase II has only one half as many chromosomes and no paired bivalents.

(d)   Mitotic telophase yields two diploid daughter nuclei; meiotic telophase II yields four haploid daughter nuclei.

(e)   Chromosomes of each bivalent have begun to separate and chiasmata have become visible by diplotene, but not by pachytene.

18-4. (a)   Yes. Each centromere is duplicated and one of each is passed to each of the daughter cells resulting from mitosis. All somatic cells arise from the zygote by mitosis, and the products of mitosis are always equal.

(b)   No. Centromeres separate at the first meiotic division, and independent assortment will randomize the segregation of maternal and paternal centromeres. Each first-division daughter cell will receive a maternal or a paternal centromere for each homologous pair.

18-5. (a)   2X   (c)   2X   (e)   4X   (g)   2X

(b) $\frac{1}{2}$ X   (d) $\frac{1}{2}$ X   (f)   2X

18-6. If the bivalent containing the two members of the chromosome 13 pair failed to separate at anaphase I of meiosis, this nondisjunction event would lead to the formation of some gametes containing two copies of chromosome 13 (as well as some gametes containing no copies of chromosome 13). If a gamete containing two copies of chromosome 13 were to fertilize a normal gamete (containing one copy of chromosome 13), the resulting zygote would have an extra copy of chromosome 13 (i.e., it would contain three copies of chromosome 13 instead of the normal two).

In a similar fashion, nondisjunction of the XY chromosome pair during anaphase I of male gamete formation, or nondisjunction of the XX chromosome pair during anaphase I of female gamete formation, could lead to the production of gametes containing neither an X chromosome nor a Y chromosome. If a sperm containing neither an X chromosome nor a Y chromosome were to fertilize a normal egg (which contains a single X chromosome), the result would be a zygote with only a single X chromosome. Similarly, if an egg containing no X chromosome were fertilized by a normal sperm containing an X chromosome, the result would be a zygote with only a single X chromosome.

18-7. (a)   The four possible genotypes of the offspring, in their respective ratios, are 1 *YY*, 2 *Yy*, and 1 *yy*. Because all *YY* and *Yy* offspring will be yellow and only *yy* offspring will be green, offspring with yellow and green seeds will appear in the ratio 3:1.

(b)   Both parents are heterozygous for both factors, so the number of gametes is 2^2, or 4. A 4 × 4 matrix has 16 outcomes for offspring, although not all of the 16 are different. In general, the number of different gametes is 2^n, where n is the number of heterozygous allelic pairs.

(c)   The Punnett square assumes that the frequency of occurrence of a given genotype among the boxes represents the frequency of occurrence of that genotype among the progeny of the genetic cross represented by the Punnett square. This assumption is only correct if all possible combinations are equally likely, which in turn depends on independent assortment of the alleles for seed color and seed shape.

(d)   The completed matrix will have 16 genotypes, nine of which are different from one another. In their respective ratios, these are 1 *YYRR*, 2 *YYRr*, 1 *YYrr*, 2 *YyRR*, 4 *YrRr*, 2 *Yyrr*, 1 *yyRR*, 2 *yyRr*, and 1 *yyrr*.

(e)   The phenotypes are yellow and round, yellow and wrinkled, green and round, and green and wrinkled. These occur in the ratio 9:3:3:1.

18-8.   (a)   The four genes are arranged as follows:

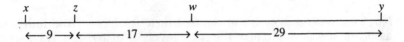

(b)   According to the map, the distance between genes *z* and *y* should be 17 + 29 = 46 map units, yet the data give a value of only 44% recombinants between these two genes. The observed value is slightly lower than expected because two recombination events (crossovers) can occasionally occur between the two genes, thereby reestablishing the parental combination. The farther apart two genes are located, the more likely that two crossover events will occur between them.

Also note that genes x and y are separated by 9 + 17 + 29 = 55 map units, yet the recombination frequency between them is only 50%. In fact, the percentage of recombinants can never exceed 50% because for genes that are assorting completely independently, an equal number of recombinants and parental combinations would be expected (i.e., a 50% recombination frequency). This will be observed if two genes are either on different chromosomes, or are located so far apart on the same chromosome that at least one crossover event is almost certain to occur between them. Since recombination frequencies cannot exceed 50%, genetic mapping is most accurate when recombination frequencies are measured on genes located relatively close to one another.

18-9.   (a)   RecBCD binds to double-strand breaks in chromosomal DNA and its helicase activity then unwinds the DNA double helix, thereby creating single-stranded loops. The nuclease activity of RecBCD cleaves one of the looped DNA strands, creating a free, single-stranded DNA end. RecA then catalyzes a "strand invasion" reaction in which this free, single-stranded DNA end invades and displaces one of the two strands of an homologous DNA double helix (step 2 in Figure 18-25 in the textbook).

(b)   While unwinding the DNA double helix with its helicase activity, the RecBCD protein moves along the DNA until it encounters a CHI site. It then stops at the CHI site and cleaves one of the DNA strands, thereby setting the stage for the RecA-catalyzed strand invasion step. Hence recombination preferentially occurs at CHI sites in cells containing RecBCD.

18-10.   In principle, the most straightforward approach would be to use the table of the genetic code (see Figure 19-8 in the textbook) to determine a DNA sequence that would code for the amino acid sequence of the known polypeptide. The researcher could then synthesize an artificial gene in a test tube, adding appropriate bacterial control sequences at either end. This gene could be inserted into a bacterial plasmid by recombinant DNA techniques and the recombinant plasmid inserted into a bacterium for cloning. The gene would be replicated with each replication of the plasmid, and, with luck, the bacteria would transcribe and translate the gene to produce the desired protein.

Alternatively, an intron-less version of the natural gene could be cloned from a cDNA library prepared using mRNA from human gonadal tissue (which would presumably contain mRNA for the desired gene). Using a procedure like the one outlined in Figure 18-27 in the textbook, the researcher could prepare a large number of bacterial clones carrying DNA from the library and then screen for the desired clone. One possible screening method would involve a nucleic acid probe, probably consisting of a variety of oligomers with sequences likely to be complementary to parts of the gene. Another method would use radioactively labeled antibody specific for the protein of interest. (Because the protein has been sequenced, we can probably assume that the researcher can obtain the small amount of pure protein needed for stimulating antibody production in a laboratory animal.) The latter screening method would be dependent on the successful production of the protein by the bacteria.

19

Gene Expression I: The Genetic Code and Transcription

19-1. (a) Degenerate: A given amino acid can be specified by more than one codon; G.

(b) Unambiguous: A given codon specifies a single amino acid; G, M (Exception: In cells containing suppressor tRNA, the code is ambiguous.)

(c) Triplet: There are three nucleotides in every codon; G.

(d) Universal: The same code is always used; G (except for a few variations in mitochondria, certain bacteria, and certain protozoa); M.

(e) Nonoverlapping: Adjacent codons do not share common nucleotides; G, M.

19-2. (a) 5'-AAA(or G) AGUCCAUCACUUAAUGCN-3' (where N is any nucleotide)

(b) 5'-AAA(or G) GUCCAUCACUUAAUGGCN-3'

(c) In terms of the mRNA, we know that an A was deleted and a G was inserted; in terms of the DNA template strand, these changes would correspond to a deleted T and an inserted C. However, we don't know *exactly* where along the DNA strand these changes occurred. Speaking again in terms of the mRNA, if the Lys codon at the 5' end of the original sequence was AAA, making the first four nucleotides AAAA, any of these A nucleotides could have been deleted; on the other hand, if the Lys codon was AAG, the deleted A must have been the next one in the sequence. The new G could have been inserted on either side of the G that is the third nucleotide from the 3' end.

19-3. (a) Not OK   (d) Not OK   (g) Not OK
(b) OK   (e) OK   (h) OK
(c) Not OK   (f) OK   (i) OK

19-4. (a) Single-nucleotide substitution:

(i) One (or possibly no) amino acid change.

(ii) Change of 1, 2, or 3 amino acids (depending on wobble and degeneracy of code).

(b)   Single-nucleotide deletion:

    (i)   Frameshift, with likely change of all downstream amino acids.

    (ii)   Loss of one amino acid, change of none, 1, or 2 others.

(c)   Deletion of three consecutive nucleotides:

    (i)   Loss of one amino acid, possible change in another, depending on reading frame.

    (ii)   Loss of three amino acids, change of none, 1, or 2 others.

19-5.   (a)   These data suggest that the promoter for the 5S RNA gene is located between nucleotides 47 and 83, since deletions that include nucleotides within this stretch prevent RNA polymerase III from transcribing the gene. Note that this type of promoter lies *within* the gene—that is, it is located downstream from the startpoint.

    (b)   Promoters for RNA polymerase I typically lie in the region between –45 and +20 (see Figure 19-13a in the textbook). Therefore, if a gene transcribed by RNA polymerase II had deletions in the same locations as the first four entries listed in the table (–45 through –1, or +1 through +47, or +10 through +47, or +10 through +63), transcription would not be expected to take place because some of the nucleotides between –45 and +20 would be missing. Deletions corresponding to the last two entries in the table (+80 through +123 and +83 through +123) would not be expected to affect promotion.

    (c)   Promoters for RNA polymerase II are of two general types. (1) TATA-driven promoters contain an Inr sequence and a TATA box, with or without an associated BRE immediately upstream from the TATA box. Therefore, if a gene transcribed by RNA polymerase II had deletions in the same location as the first two entries in the table (–45 through –1 or +1 through +47), transcription would not be expected to occur because either the TATA box or the Inr sequence would be disrupted. Deletions corresponding to the last four entries in the table would not be expected to affect promotion because they are all downstream of BRE, the TATA box, and Inr. (2) DPE-driven promoters contain Inr sequences near the startpoint and DPE sequences about 30 nucleotides downstream from Inr. Therefore, deletions corresponding to the first two entries in the table (–45 through –1 or +1 through +47) would both fail to be transcribed because the Inr sequence would be disrupted. In addition, deletions corresponding to the second, third, and fourth entries in the table (+1 through +47, +10 through +47, and +10 through +63) would fail to be transcribed because the DPE sequence would be disrupted. Deletions corresponding to the last two in the table would not be expected to affect promotion because they are all downstream from Inr and DPE.

19-6.   (a)   B, I   (g)   none
    (b)   B, I, II, III   (h)   B, I, II, III
    (c)   I, II, III   (i)   II
    (d)   B   (j)   B, I, II (Only RNA polymerase III promoters may lie entirely downstream of the transcriptional startpoint.)

(e)   B, I, III   (k)   B

(f)   I, II, III

19-7. (a) rRNA: Cleavage of large precursor; degradation of transcribed spacers.

tRNA: Removal of leader sequence at 5′ end; addition of CCA sequence at 3′ end (if not already present); methylation of bases; removal of introns.

mRNA: Elimination of introns; addition of poly (A) tail on 3′ end; capping of 5′ end.

(b) rRNA: Generation of several molecules from one large precursor.

tRNA: Addition of CCA sequence at 3′ end if not already present.

mRNA: Capping of 5′ end; addition of poly (A) tail.

(c) rRNA: Generation of several molecules from one large precursor; degradation of transcribed spacers.

tRNA: Addition of CCA sequence if necessary; methylation of bases; cleavage from a larger precursor.

mRNA: None in common.

19-8. (a) R; snRNA are the small (nuclear) RNA molecules that are incorporated into the spliceosome. In addition to a structural role in spliceosome assembly, snRNA molecules recognize (by base pairing) the splice sites on the primary RNA transcript and probably play a major role in catalyzing the splicing reaction.

(b) PR; spliceosomes, the large molecular complexes where splicing occurs, consist of snRNPs (see part c) and additional proteins.

(c) PR; a variety of snRNPs, each consisting of one or two molecules of snRNA plus proteins, assemble with other proteins to form a spliceosome.

(d) R; splice sites are the specific nucleotide sequences in the primary RNA transcript at the junctions between exons and introns. The spliceosome recognizes these sequences and cuts and splices at particular points within them.

(e) R; when splicing occurs, the RNA of each intron leaves the transcript in the form of a lariat, which is subsequently degraded.

19-9. (a) Rifamycin interferes with initiation because it prevents formation of the first phosphodiester bond.

(b) Actinomycin D interrupts elongation because it blocks the DNA.

(c) Actinomycin D, because rifamycin is specific for bacterial polymerase.

(d) Rifamycin, because it inhibits initiation but does not block elongation.

19-10. (a)   Codons: UUU, UUC, UCU, CUU, CCU, CUC, UCC, CCC.

           Amino acids: phenylalanine (UUU, UUC), serine (UCU, UCC), leucine (CUU, CUC), proline (CCU, CCC).

   (b)   Incubation A: 8 of each.

           Incubation B: 27 UUU; 9 each of UUC, UCU, and CUU; 3 each of CCU, CUC, and UCC; and 1 CCC.

   (c)   Incubation A: 16 of each.

           Incubation B: 36 phenylalanine; 12 each of serine and leucine; 4 proline.

   (d)   It is possible to distinguish between classes of codons that differ in their base composition and to determine the base compositions of the codons that code for various amino acids.

   (e)   Yes, the data of parts (b) and (c) for incubation B should allow this to be deduced.

   (f)   No.

   (g)   Synthesize the relevant codons and test them in the rRNA-binding assay of Nirenberg.

19-11. (a)   At least one intron must be present in gene "X". This intron contains a restriction site for *Hae*III, explaining why digestion with *Hae*III yields an extra fragment when the gene rather than the cDNA is cleaved with *Hae*III. It is possible that more than one intron is present in the gene, because the additional introns might not contain restriction sites for *Hae*III, and hence their presence could not be detected by digesting the DNA with *Hae*III.

   (b)   Although two extra fragments are produced when the gene rather than the cDNA is cleaved with *Hae*III, this does not necessarily mean that two introns are present in gene "Y" because the two additional *Hae*III sites could both be located in the same intron. Hence you can only conclude that *at least* one intron is present.

   (c)   Since no extra fragments are produced when the gene rather than the cDNA is cleaved with *Hae*III, this result does not provide any evidence for the existence of an intron within gene "Z". However, it is possible that one or more introns are present in the gene, because these introns might not contain restriction sites for *Hae*III, and hence their presence could not be detected by digesting the DNA with *Hae*III.

20

Gene Expression II:
Protein Synthesis and Sorting

20-1. (a) 5' UUAAU<u>AUG</u>UGCUACUUCGAACACUGUCCCAAAGG<u>UUAG</u><u>UAA</u>UU 3'

3' AAUUAUACACGAUGAAGCUUGUGACAGGGUUUCCAAUCAUUAA 5'

(b) Only the first of the two RNA sequences has an initiation codon (AUG, under-scored above) in the correct (5' → 3') direction, so it must be the messenger. Notice also that the same RNA sequence has two stop codons in the correct reading frame near the 3' end (UAG, UAA, also underscored).

(c) The amino acid sequence of vasopressin is therefore:

AUG UGC UAC UUC GAA CAC UGU CCC AAA GGU UAG UAA

Met—Cys—Tyr—Phe—Glu—His—Cys—Pro—Lys—Gly—(stop)

(0)   1   2   3   4   5   6   7   8   9

(d) The methionine at the N-terminal end is apparently cleaved to generate the mature polypeptide.

(e) Oxytocin differs from vasopressin in positions 3 (Ile for Phe), 5 (Asp for His), and 8 (Leu for Lys). The first two amino acid changes can be accomplished by one-base changes in the mRNA (AUC instead of UUC, GAC instead of CAC), but the third requires a two-base change (UUA or CUA instead of AAA). The most conservative DNA sequence would therefore be:

3' AATTATACACGATGTAGCTTCTGACAGGGAATCCAATCATTAA 5'

The similarity between the coding sequences for the two hormones suggests an evolutionary relationship between them. Either one evolved from the other, or both evolved from a common ancestral sequence.

20-2. (a) Arg: AGA

Ile: AUA

Lys: AAA

Stop: UAA

Thr: ACA

Ser: UCA

(b)
UAU (or C)	Tyr	1st position
UCA	Ser	2nd position
UUA	Leu	2nd position
CAA	Gln	3rd position
AAA	Lys	3rd position
GAA	Glu	3rd position

20-3. In the initiation of eukaryotic translation, the first few binding steps occur in an order different from that in prokaryotic initiation. In the prokaryotic case, initiation factors and GTP attach to a small ribosomal subunit before the initiator tRNA (carrying fMet) arrives. In the eukaryotic case, an initiation factor binds to GTP and to the initiator tRNA (carrying Met) before any of them bind to the small ribosomal subunit. The way in which the small ribosomal subunit arrives at the start codon on the mRNA also differs. In prokaryotes, the ribosome binds to a ribosome-binding site (Shine-Dalgarno sequence) on the mRNA, which is located near the start codon. In eukaryotes, the small ribosomal subunit-tRNA complex initially binds to the mRNA at its 5' cap and then moves down the mRNA until it reaches the start codon, where the tRNA anticodon base-pairs. Only then does the large ribosomal subunit join the complex.

20-4. (a) P (c) B (e) B (g) B (i) B
      (b) B (d) N (f) N (h) B

20-5. (a) Because it looks like an aminoacyl tRNA and contains an α amino acid, puromycin can bind to the ribosome and form a peptide bond with the carboxyl group of the growing peptide chain. The product, peptidyl puromycin, then dissociates from the ribosome.

      (b) To the carboxyl end, because it has an amino group.

      (c) To the A site, because it resembles an aminoacyl tRNA.

      (d) Equally effective, because it interferes with the polypeptide elongation stage of protein synthesis, which is very similar in prokaryotic and eukaryotic cells.

20-6. (a) Rambomycin inhibits translocation of peptidyl tRNA from the A site to the P site, with concomitant movement of the mRNA. Inhibition of any earlier step would prevent formation of the first peptide bond.

      (b) Both products will remain as peptidyl tRNA because there is no stop signal present to activate a release factor.

20-7. A mutant tRNA with a *four*-nucleotide anticodon could suppress a frameshift mutation consisting of a one-nucleotide insertion nearby. A mutant tRNA with a *two*-nucleotide anticodon could suppress a frameshift mutation consisting of a one-nucleotide deletion.

20-8. (a) BiP contains a KDEL sequence, causing it to be retained in the ER.

      (b) If BiP lost the ability to bind to hydrophobic amino acids, it could no longer bind to unfolded and misfolded polypeptides. The exposed hydrophobic regions of these unfolded and misfolded polypeptides would cause them to aggregate with each other, forming insoluble deposits that can trigger disruptions in cell function and may even lead to cell death.

(c)   Although it is possible that polypeptide X might be stable in the absence of polypeptide Y, the different polypeptides that make up a multisubunit protein must often interact with each other in order to fold into their proper three-dimensional conformation. Thus, in the absence of polypeptide Y, polypeptide X might remain bound to BiP because it cannot fold properly in the absence of polypeptide Y. Polypeptides that repeatedly fail to fold properly are eventually shuttled back across the ER membrane to the cytoplasm, where they are degraded.

20-9.   (a)   In the absence of SRP and ER membranes, a prolactin *precursor* molecule is synthesized. This precursor, called preprolactin, consists of the normal prolactin sequence (199 amino acids) plus a 28 amino acid ER signal sequence at the N-terminus.

         (b)   SRP binds to the ER signal sequence of the newly forming polypeptide chain and halts protein synthesis, thereby preventing the chain from growing beyond 70 amino acids. Inside cells, this blockage is normally maintained until the SRP binds the ribosome-mRNA complex to the ER. Without this blockage, a complete preprolactin chain might be produced by cells and released into the cytosol, rather than being transported into the ER lumen for secretion from the cell.

         (c)   In the presence of ER membrane vesicles, the SRP binds the mRNA-ribosome complex to the ER, where signal peptidase cleaves the ER signal sequence from the newly forming prolactin chain as it moves across the ER membrane. Removal of the ER signal sequence reduces the length of the final polypeptide from 227 amino acids to 199 amino acids. The final polypeptide chain is released into the lumen of the ER vesicles.

20-10.   (a)   B   (e)   Nu
            (b)   M   (f)   M
            (c)   M   (g)   B
            (d)   M   (h)   Nu

21

The Regulation of
Gene Expression

21-1. Activity of the z gene

	Presence of Lactose	Absence of Lactose	Explanation
(a)	+	−	Wild-type; inducible system
(b)	−	−	Superrepressor cannot recognize inducer
(c)	+	+	Operator cannot bind repressor
(d)	+	+	Repressor not made; system always "on"
(e)	+	+	Operator cannot bind repressor
(f)	−	−	Promoter cannot bind RNA polymerase
(g)	−	−	Glucose reduces CRP binding

21-2. (a) Genes A and B are structural genes, probably coding for enzymes involved in the pathway for ethanol synthesis. Genetic locus C is the operator, and D is the gene encoding the repressor.

   (b) (i) No synthesis of enzyme A, inducible for enzyme B; cannot produce ethanol

   (ii) Inducible for both enzymes; wild-type phenotype

   (iii) Inducible for both enzymes; wild-type phenotype

   (iv) Inducible for enzyme A, constitutive for enzyme B; wild-type phenotype

21-3. (a) C   (c) X   (e) C   (g) C   (i) I
       (b) C   (d) I   (f) I   (h) C

21-4. (a) Their mode of action is to terminate transcription prematurely to varying extents under different conditions of concentration.

   (b) Signals termination, causing dissociation of the RNA polymerase from the template.

   (c) The transcriptional activity of an RNA polymerase molecule is influenced by changes in mRNA structure induced by a ribosome that is translating that mRNA.

(d)   It ensures that the ribosome does indeed "tailgate" the RNA polymerase, allowing the necessary proximity of the ribosome to the RNA polymerase molecule.

(e)   Has amino acid X present at more than one position.

(f)   The model requires sensitivity of transcription to aminoacyl tRNA availability and therefore to the presence of a specific amino acid in the leader peptide.

(g)   The close coupling of transcription and translation on which the model depends is not possible in eukaryotes because of the nuclear envelope.

21-5.   (a)   B, G   (c)   N   (e)   L   (g)   L
      (b)   L   (d)   G   (f)   N   (h)   N

21-6.   (a)   Because the number of genes initially estimated by mutation studies was roughly the same as the total number of chromosomal bands (5000), it seemed logical to infer that each band might contain only one gene. However, 5000 proteins seems barely adequate to code for all the structural and functional needs of a complex, multicellular eukaryotic organism, given that *E. coli* has about 4300 genes. DNA sequencing studies have revealed the presence of more than 13,000 genes (see Table 16-2), suggesting that the early mutational studies significantly underestimated the total number of genes present.

(b) $(2 \times 10^8)\ (0.75)/5000 =$ **30,000 base pairs per band.**

(c)   50,000/110 = 455 amino acids per protein.

455 × 3 = 1365 base pairs per protein.

1365/30,000 = 0.045 = **4.5%.**

(d)   The discrepancy could result from (1) the presence of some DNA in each band that is not transcribed; (2) transcription of DNA into some RNA sequences that do not form part of the coding message; and (3) the presence of more than one gene in some bands. It appears likely that all three explanations are involved.

21-7.   (a)   True. Enhancers have been shown to increase rates of transcription initiation even when located up to tens of thousands of base pairs from the promoter that they affect.

(b)   True. In a few cases, enhancers located within genes have been identified and observed to increase transcription levels. For example, the introns of some immunoglobulin genes contain enhancers.

(c)   False. Regardless of an enhancer's position, it cannot increase the rate of transcription of a gene unless the gene has a core promoter. The core promoter region provides the binding site for TFIID and RNA polymerase.

(d)   True. Enhancers appear to act by forming a loop of DNA as the activators bound to the enhancer bind, via different domains, to coactivators which are subunits of TFIID, which in turn binds to the promoter.

(e)   False. Enhancers have not been shown to play a role in DNA splicing.

21-8. (a) $(500)(13,000)/(2.7 \times 10^9) = 0.0024 = \mathbf{0.24\%}$.

(b) $(2.7 \times 10^9) + (5 \times 10^5)(13,000) = \mathbf{9.2 \times 10^9}$ **base pairs** per amplified haploid genome.

Ribosomal genes represent $6.5/9.2 \times 100\%$, or **70%**, of the amplified genome!

(c) Because amplification is 1000-fold, the unamplified genome would presumably require 1000 times longer to synthesize the required number of ribosomes. Therefore 2000 months, or about **167 years**, would be needed (which would make female frogs *very* old mothers!).

(d) When the desired gene product is an RNA, all the molecules the cell needs must be transcribed directly from the DNA. When the desired product is a protein, its mRNA can be translated repeatedly, so there is an additional stage of amplification at the mRNA level.

21-9. (a) Testosterone and hydrocortisone bind to different hormone receptors present in liver cells. The binding of testosterone allows its receptor to bind to testosterone response elements in DNA, whereas the binding of hydrocortisone allows its receptor to bind to glucocorticoid response elements in DNA. The gene coding for α2-microglobulin is associated with a testosterone response element, whereas the gene coding for tyrosine aminotransferase is associated with a glucocorticoid response element. As a result, testosterone and hydrocortisone activate the transcription of the genes coding for α2-microglobulin and tyrosine aminotransferase, respectively.

(b) Steroid hormones such as testosterone and hydrocortisone bind to receptors that can activate gene transcription, leading to the production of mRNAs that are then translated into new polypeptide chains. Therefore, inhibitors of either mRNA synthesis (e.g., α-amanitin) or protein synthesis (e.g., puromycin) would be expected to block the ability of these two hormones to stimulate the production of α2-microglobulin and tyrosine aminotransferase, respectively.

(c) The zinc-finger domain is responsible for the ability of hormone receptors to recognize and bind to the specific DNA sequences that make up their corresponding DNA response elements. Therefore, switching the zinc-finger domain between the testosterone and glucocorticoid receptors would be expected to switch the DNA-binding specificities of the two receptors. As a result, the testosterone receptor would bind to glucocorticoid response elements, and the hydrocortisone receptor would bind to testosterone response elements. The binding of testosterone to its receptor would now be expected to increase the production of tyrosine aminotransferase, and the binding of hydrocortisone to its receptor would be expected to increase the production of α2-microglobulin.

(d) Testosterone would increase the production of both α2-microglobulin and tyrosine aminotransferase, whereas hydrocortisone would increase the production of neither of these proteins.

21-10. (a) False. Homeotic genes play key roles in early *Drosophila* development by directing synthesis of transcription factors that coordinate the development of fundamental characteristics such as body shape and appendage location.

Furthermore, homeotic proteins contain helix-turn-helix domains, not zinc fingers.

(b)   False. Mutations in homeotic genes are not necessarily fatal for *Drosophila*, but they often lead to very dramatic abnormalities, such as legs sprouting from the fly's head.

(c)   True. Homeotic proteins form a family of widely active transcription factors, each of which may bind to and control expression of hundreds of other genes in a *Drosophila* embryo.

(d)   False. Homeotic genes, which are found in a wide range of eukaryotes, are not identical, but they are related by virtue of containing a highly conserved 180 nucleotide sequence known as the homeobox.

(e)   False. As transcription factors, homeotic proteins affect gene expression at the level of RNA synthesis.

21-11. (a)   One possibility is that the missing stretch of amino acids contains the lysine to which ubiquitin is normally attached when mitotic cyclin is being targeted for destruction. A second possibility is that the missing segment contains a destruction box whose detection by an appropriate recognition protein normally targets mitotic cyclin for ubiquitination. Finally, the removal of a stretch of amino acids could change the conformation of mitotic cyclin so that it can no longer serve as a substrate for the ubiquitinating enzyme complex.

(b)   The amino acid sequence of the ubiquitination site of normal mitotic cyclin could be analyzed to determine whether the lysine residue that serves as an attachment site for ubiquitin is missing in the mutant cyclin. Alternatively, recombination DNA techniques could be used to insert the missing stretch of nine amino acids into other proteins to see whether it can serve as a destruction box that targets these proteins for degradation. If the preceding kinds of experi-ments fail to provide evidence that the deleted stretch of amino acids serves as either a ubiquitination site or a destruction box, then it is possible that that amino acid deletion is simply exerting its effects by altering protein conformation, although this hypothesis would be difficult to prove directly.

(c)   A mutation in a recognition protein that specifically binds to a destruction box present in mitotic cyclin would explain such observations.

21-12. (a)   The total mRNA population of one tissue could be radioactively labeled and hybridized against total mouse DNA (or, for greater sensitivity, against the nonrepeated DNA component of the mouse genome) in the presence and absence of mRNA from the other tissue. The extent to which radioactive RNA from one tissue could be "competed out" by nonradioactive species from the other tissue measures the fraction of common sequences. Alternatively, DNA microarrays could be used to compare the mRNAs produced in the two tissues.

(b)   Selective processing of nuclear transcripts and/or selective translocation of RNA to the cytoplasm could also explain the data. To distinguish between these possibilities, nuclear RNA preparations from the two tissues could be tested for the presence or absence of common sequences by the same hybridization techniques used with the cytoplasmic RNA.

(c)   Translational control would seem likely.

21-13. As part of the normal process of female gametogenesis, frog eggs produce and store large amounts of mRNA. After an egg is fertilized by a sperm cell, these mRNAs are translated into the proteins that are required during the early stages of embryonic development. Since these mRNAs are already present, inhibitors of RNA synthesis have relatively little affect on either protein synthesis or early development. Later in development, genes coding for proteins required during the later stages of embryonic development begin to be transcribed into mRNAs that are then translated into polypeptides. At this point, blocking transcription with inhibitors of RNA synthesis will prevent these essential proteins from being produced.

22

Cytoskeletal Systems

22-1   (a)   MF   (c)   MT   (e)   IF   (g)   IF   (i)   MF, MT

         (b)   MT   (d)   MF   (f)   N   (h)   MF, MT, IF   (j)   MT, IF

22-2.   (a)   F. The *rate* of loss is greater than the *rate* of addition at the minus end over time. There is no absolute absence of addition at the minus end.

         (b)   F; hydrolysis of the ATP bound to actin and the GTP bound to tubulin usually occurs during monomer polymerization but the polymerization process still occurs even if the ATP or GTP is replaced by a nonhydrolyzable analogue.

         (c)   F; the statement is true for microtubules and microfilaments, but there is no evidence for dynamic assembly and disassembly of intermediate filaments.

         (d)   F; cytochalasin D prevents actin polymerization and therefore inhibits cytokinesis but not chromosome movement, because actin is involved in the former but not the latter.

         (e)   F; algae are eukaryotes and therefore possess the same kinds of cytoskeletal components as other eukaryotic cells.

         (f)   T

         (g)   F. Most do, but not all. In addition, in ciliated or flagellated cells, the minus ends of microtubules can be anchored at the basal body, which also serves as a microtubule-organizing center.

         (h)   M; the statement is true if the monomer concentration is above the overall critical concentration but false otherwise.

22-3.   (a)   Pigment granule dispersal is a microtubule-dependent process.

         (b)   The centrosome serves as a microtubule-organizing center in vivo, and all of the microtubules radiating from the centrosome apparently have the same polarity.

         (c)   Although each kind of muscle cell has actin molecules that are specific for that muscle type, the differences between the several species of actin are apparently too small to have any discernible effect on actin polymerization, structure, or function.

(d)   The extracts appear to contain structures that are functionally equivalent to centrosomes (as evidenced by the presence of pericentrin), which nucleate microtubule growth.

22-4.   The overall critical concentration should be reduced when tubulin polymerizes in the presence of the centrosome preparation. If the minus end cannot disassemble, the overall critical concentration should now be equivalent to the lower critical concentration of the plus end.

22-5.   (a)   It would be reasonable to postulate that the G-actin is kept from polymerization prior to contact with an egg cell by being complexed one-to-one with profilin. The profilin-actin complex, unlike free G-actin, would not be able to initiate polymerization. To test this hypothesis, profilin could be isolated and added to a mixture of G-actin monomers in vitro to determine whether the profilin can bind to actin and prevent the initiation of microfilament assembly under conditions that would normally be optimal for polymerization.

(b)   Because the pH rise precedes actin polymerization, it would be reasonable to postulate that the change in pH destabilizes the profilin-actin complex, causing dissociation into free actin, which can then assemble into F-actin. To test this hypothesis, one could vary the pH of the mixture of profilin and actin and determine whether profilin-actin complexes are present at high, but not low, pH. One could also examine the effects of pH on actin polymerization in the presence of profilin in vitro.

22-6.   Cytochalasin D prevents addition of new monomers to existing filaments, effectively capping them. In situations in which the concentration of G-actin in the cytosol is below the critical concentration for net addition of subunits to the minus end of the capped filaments, loss of monomers at the minus end of existing filaments eventually results in their shortening. This occurs despite the pool of available G-actin in the cytosol.

22-7.   (a)   Gelsolin severs microfilaments, resulting in loss of tension-generating elements, so wrinkling should decrease.

(b)   Intermediate filaments act as structural scaffolding within cells. The absence of a major component of the intermediate filament network shoul result in loss of rigidity. Because the internal structural elements of the cell themselves can no longer resist the tension generated by the cells, the rubber must bear more of the stress generated by the cells. This should result in more wrinkling.

(c)   Taxol stabilizes microtubules, which are thought to bear compressive loads. Now the cells should be better able to resist their own tension, and the wrinkling of the rubber should be less pronounced.

(d)   Colchicine would cause the loss of microtubules, resulting in the opposite effects as in (c), i.e., more wrinkling should result. After washing out the drug, microtubules will regrow, and the cells should be better able to resist tension internally. This should eventually result in the reduction of wrinkling of the rubber over time.

23

Cellular Movement: Motility and Contractility

23-1. During anaphase, microtubules with opposite orientations overlap in the central region of the spindle. One possibility is that the Cho1/MKLP1 kinesins are involved in the sliding apart of these microtubules that is known to occur in anaphase B. Thus we might expect defects to arise during the late stages of mitosis. Actually, Cho1/MKLP1 is thought to be involved not only in the events of late anaphase, but also in allowing the completion of cytokinesis. The latter role of this protein is not well understood.

23-2. (a) The nonmotility of the sperm cells (and hence the sterility of the individual) is due to a structural defect in the outer dynein arms of the sperm tail axoneme, causing the sperm tail (or flagellum) to be nonfunctional. (In addition to the defect underlying Kartagener's triad, there are several other defects that also result in sperm nonmotility and hence male infertility. These include a defect in, or lack of, any of the following: both dynein arms, the inner arm only, the radial spoke heads, or one or both microtubules of the central pair.)

(b) The same structural defects in the sperm tail are also likely to affect the cilia responsible for sweeping mucus and foreign matter out of the lungs and sinuses.

23-3. (a) A, H (c) A, H, I, R (e) A, H, I (g) A
(b) R (d) I (f) H

23-4 (a) At a sarcomere length of 3.2 μm, the length of the A band is 1.6 μm, and the length of the I band is also 1.6 μm, with 0.8 μm on either side of the Z line. During contraction of the sarcomere to 2.0 μm, the length of the A band remains fixed at 1.6 μm, while the length of the I band decreases from 1.6 μm to 0.4 μm.

(b) The H zone corresponds to that portion of the thick filament length that is not overlapped by thin filaments. (It is, in fact, the lack of interdigitated thin filaments that gives the H zone its lighter density and hence its German name.)

(c) The distance from the Z line to the edge of the H zone represents the length of the thin filament and remains constant during contraction.

23-5. (a) Rigor results from a failure to break the cross-bridges that link thick filaments to thin filaments in the contraction cycle. In the living cell, detachment occurs upon binding of the next molecule of ATP. After death, however, the supply of cellular ATP is quickly depleted and cannot be restored. This has two consequences: (1) ATP is not available to cause cross-bridge detachment; and (2) in the absence of ATP, calcium cannnot be pumped into the sarcoplasmic reticulum. Calcium therefore accumulates in the cytosol and promotes attachment of cross-bridges to actin. The net result is an accumulation of cross-bridges that lock the muscle filaments together and give the corpse its characteristic stiffness.

(b)   While running to class; the concentration of ATP in the cell would presumably be lower and cross-bridge formation would probably occur more quickly upon death than if you were just sitting in lecture.

(c)   Addition of ATP will have a relaxing effect, because it will allow cross-bridges to break and filaments to detach.

23-6.   (a)   The cross-bridges will be dissociated from the thin filaments, but will not be "recocked" to the conformation normally seen in resting muscle, because AMPPCP cannot be hydrolyzed to ADP and P$_i$. Your diagram should look like that after step 2 in Figure 23-18 in the textbook, except that AMPPCP is bound to the myosin head instead of ATP.

(b)   Yes, because as a structural analogue, AMPPCP is presumably a competitive inhibitor of the ATPase and should be displaced from the binding site if the intracellular ATP concentration is elevated to a high enough level.

(c)   All other ATP-dependent processes are likely to be inhibited, including uptake of calcium ions by the sarcoplasmic reticulum.

23-7.   The myosin thick filaments are assembled with the rodlike tail domain in the center of the filament and the globular heads located at the ends of the filament pointing away from the center. Therefore, the globular myosin heads at each end of the thick filament have opposite polarities. This results in the movement of the actin filaments in opposite directions so that actin filaments from each Z line are drawn to the center of the sarcomere.

23-8.   (a)   If the fibroblast has been detached, there is no resistance to retrograde flow (i.e., the "clutch" is not engaged). Therefore retrograde flow should balance subunit addition and fluorescent spots should move rearward.

(b)   Cytochalasin treatment will prevent the addition of new actin monomers at the leading edge. Spots located on existing filaments should still translocate rearward. However, no new filaments will replace them.

(c)   Such a cell has its "clutch" engaged, so retrograde flow will be resisted relative to sites of attachment to the substrate on the underside of the cell, and spots should not move rearward relative to these sites of attachment.

23-9.   (a)   One could employ compounds that selectively disrupt microfilaments (such as cytochalasin D) or microtubules (such as colchicine) to inhibit movement of the melanophores.

(b)   In order to get contraction, there must be actin filaments with opposite polarities present within the stress fiber.

(c)   In order to get the wavelength motion seen in the beating of sperm flagella, some dynein molecules must be exerting force while those on the opposite side of the tail must be inactive. If dynein molecules were simultaneously exerting force on all sides, the sperm tail would be locked in a rigid state, unable to move at all.

(d)   *Listeria* moves within infected cells via "comet tails" of polymerized actin. Since latrunculin prevents actin polymerization by sequestering actin monomers, *Listeria* translocation—and therefore the further spread of *Listeria* infection—will be blocked.

23-10. (a)   A resting fibroblast is not in a state in which most of its Rho is activated, so we would expect the introduction of a dominant negative Rho to have little effect on the cell.

(b)   PDGF tends to activate Rac, not Rho, so the presence of a dominant negative Rho would be expected to have little effect on Rac activation in the fibroblast cell. The result would likely be the production of lamellipodia by the treated cell.

(c)   Constitutively activated Rho would tend to enhance the response of LPA treatment, since LPA typically activates the Rho pathway. Thus, we would expect even more stress fiber formation in the LPA-treated fibroblast than in normal cells treated with LPA.